William Rice Pryor

The Treatment of Pelvic Inflammations through the Vagina

William Rice Pryor

The Treatment of Pelvic Inflammations through the Vagina

ISBN/EAN: 9783337211509

Printed in Europe, USA, Canada, Australia, Japan

Cover: Foto ©berggeist007 / pixelio.de

More available books at **www.hansebooks.com**

THE TREATMENT

OF

PELVIC INFLAMMATIONS

THROUGH THE VAGINA

BY

WILLIAM R. PRYOR, M.D.

PROFESSOR OF GYNECOLOGY, NEW YORK POLYCLINIC; CONSULTING SURGEON,
CITY (CHARITY) HOSPITAL; VISITING SURGEON, ST. ELIZABETH
HOSPITAL, NEW YORK CITY.

WITH 110 ILLUSTRATIONS

PHILADELPHIA:
W. B. SAUNDERS
925 Walnut Street
1899

TO

FERNAND HENROTIN, M.D.,

IN

ADMIRATION OF HIS SURGICAL ABILITY

AND AS A

TOKEN OF THE WARMTH OF MY AFFECTION

THIS WORK

Is Inscribed

BY

THE AUTHOR.

PREFACE.

This little book has been written at the request of the gentlemen who have attended my lectures in the New York Polyclinic, and is but an elaboration of what I have said and done before my audiences.

There exists the utmost confusion in the profession regarding the most successful methods of treating pelvic inflammations; and inasmuch as inflammatory lesions constitute the majority of all pelvic diseases, the subject is an important one: Furthermore, these cases are, as a rule, emergency cases. An attending physician cannot waste time in studying up an operation to be performed, nor can he transport his patient to a distant hospital.

I have gone into pathology only so far as will enable the operator to identify the lesions. Some stress has been laid upon the physical characteristics of the intra-peritoneal lesions as revealed by vaginal section, for it is upon such an inspection that the nature of further work must be based. It has been my endeavor, having in mind the object for which the book is written, to put down every little detail, no matter how insignificant, which might be of service.

PREFACE.

The spirit predominant throughout this book is that of an aggressive interference. Yet I have generally laid down also a palliative method of treatment for each disease, to be applied where operation cannot be done I have told what I think and do. The views and methods of others can readily be procured from the medical press. And as I write *ex cathedra*, quotations and references are unnecessary. Still, it must not be imagined that all herein contained is claimed as original. The contribution of any one of us must nowadays be small as the science progresses, and I take pleasure in acknowledging my indebtedness to Récamier, Péan, Ségond, and Pozzi in Europe, and to Gaillard Thomas, Byford, and Henrotin in America.

I have had great assistance in the pathological work from Dr. F. M. Jeffries, who has been as interested in the work as I; Mr. C. L. Remele has been most painstaking in photographing the specimens; and to Miss Fanny Elkins is due all credit for the illustrations of the operations. I can only hope that I may succeed in directing the attention of the general practitioner to a surgical treatment of the infectious pelvic diseases of women; and if I do so, he will find in these pages some hints which may be of service to him.

WILLIAM RICE PRYOR.

CONTENTS.

	PAGE
ENDOMETRITIS	17
Acute and Chronic Septic Endocervicitis	20
Acute Gonorrheal Endocervicitis	25
Chronic and Latent Gonorrheal Endocervicitis	25
Septic Endometritis	26
Puerperal Infection	34
Acute Gonorrheal Endometritis	47
Chronic Gonorrheal Endometritis	50
Tubercular Endometritis	52
PELVIC INFLAMMATION	54
SALPINGITIS	63
Acute Gonorrheal Salpingitis	63
Acute Septic Salpingitis	69
Chronic Septic and Gonorrheal Salpingitis	72
Tubercular Salpingitis	89
PELVIC PERITONITIS	90
TUBERCULAR PERITONITIS	104
INFLAMMATORY DISEASES OF THE OVARIES	106
BROAD-LIGAMENT CYST	114
ADHERENT RETROPOSITIONS	117

CONTENTS.

	PAGE
BROAD-LIGAMENT ABSCESS	121
DIFFUSE PELVIC SUPPURATION	123
ANESTHETIC	125
CURETTAGE	126
EXPLORATORY VAGINAL SECTION	136
CONSERVATIVE TREATMENT	146
Conservative Operations upon the inflamed Adnexa Uteri	152
PREPARATION OF PATIENT FOR A VAGINAL SECTION	161
VAGINAL ABLATION	163
Ablation en Masse	180
Ablation by Hemisection	187
MORCELLATION	209
VAGINO-ABDOMINAL HYSTERECTOMY IN PUERPERAL STATE	215
AFTER TREATMENT OF HYSTERECTOMY AND VAGINAL SECTION	219
ACCIDENTS AND COMPLICATIONS	223
SECONDARY HEMORRHAGE	227
INTRAVENOUS INJECTION OF NORMAL SALT SOLUTION	229
INSTRUMENTS	231
FORMULÆ	237
STERILIZATION	238
INDEX	241

ENDOMETRITIS.

General Considerations.—Believing that the phenomena of inflammation in the female genitals have the same general significance as when occurring in other parts of the body, and that pus produced by such inflammation is caused by the same pyogenic cocci as cause inflammation elsewhere, my conception of its treatment is distinctly surgical. In view of the ravages which these inflammations, when neglected, work in woman's special organs, organs important to her no more than to the community, organs so intimately associated with her mental and social as well as her physical life, I am an advocate of energetic methods of treatment. Pyosalpinx, ovarian abscess, and peritonitis rarely occur except through the medium of the uterus. The excepted cases where the intestinal tract is to blame for the pelvic lesions are so rare that they but emphasize the rule. Such being the case, the importance of properly treating endometritis is apparent.

But while we combat their diseases, the special functions of the organs must be considered, and in our endeavors to check an invasion from without we must so act as to do the least damage to that peculiar membrane, the endometrium, whose function it is to produce the decidual cell. Therefore I advocate the use of large quantities of mild antiseptics as washes, agents which by their bulk ensure cleanliness, and by their innocuousness do not damage the tissues. Conversely, I am opposed to the use of small quantities of powerful escharotic antiseptics, such as carbolic acid, zinc chlorid, etc. The greater the degree of infection within a cavity, the less the indication for strong antiseptics; for these but destroy tissue in a locality whose vitality is already damaged by inflammation, and by the slough caused they produce the most propitious culture medium for the germs not removed. Dead tissue must be cast off by suppuration. If after curettage, for example, in chronic purulent (septic) endometritis, the raw surfaces be painted with carbolic acid or even iodin, the intra-uterine packing of iodoform gauze will be followed by temperature. Why? Because the superficial slough produced by the application is not removed by the dressing, retention results, and recovery is accompanied by fever. But if after a clean curettage the uterus be irrigated with boric acid solution, the gauze dressing protects the uterus while the endometrium is in process of repair, and the convalescence is afebrile. Sterility commonly follows one method, conception can be expected after the other. That is, the function of the uterus is partially destroyed by one, while its return to a physiological state is promoted by the other.

So far as the treatment of endometritis is involved, the physician is concerned clinically with whether it is *purulent* or *non-purulent*. If purulent, applications to the endometrium are contra-indicated: if non-purulent, applications may be used. But great care must be exercised lest the innocent form be converted into the virulent type. Whenever pus is discharged from the uterus, the woman's position is just the same as that of any other patient who has a purulent discharge from some other part of the body; it is a surgical case. With our present precise methods of examination, and in view of our ability to do clean work under all circumstances, there no longer exists any excuse for employing the opium, poultice, and douche treatment of a spreading infection. The moment a physician sees pus escape from the uterus his anxiety must begin, and the instant an extension to the adnexa or peritoneum sets in, his responsibility becomes great. A woman who has once had tubal or peritoneal disease occupies an unfortunate position in society; she has before her all her life a possibly dangerous operation. For a man to sit quietly by and see this condition brought about without attempting to check the infection is to assume a responsibility before his patient and the profession which I always refuse. Be careful lest you convert the simple milky discharge of the nullipara into a purulent one by caustics; and be again careful lest you promote an extension of a purulent disease; finding the one, let it alone; detecting the other, check it instantly.

We may well ask the simple question, why do women have peritonitis so much more often than men? The answer comes with the question. If then the open infected uterus is to blame, why not attack its filth first? Here the infection began, and from this source the fire is fed. Combat it here. To me it is as rational to incise an axillary bubo and neglect the felon which causes it, as to remove acutely inflamed tubes and ignore the causative endometritis. This I can say to the man who first sees these cases, that whereas we now do many hysterectomies for pelvic inflammation, but few would require it if the uterine inflammation were first energetically and properly treated.

It must not be understood from what I have said that I believe all forms of endometritis are caused by pathogenic germs. Undoubtedly simple endometritis may be found frequently to be of purely hematogenous origin, and such a type may have the characteristics common to those forms due to germs, except the production of pus. Such an endometritis being purely local, never inducing complications so long as it maintains its type, never causing metastases, is amenable to non-surgical measures. The leucorrhea of the young girl is relieved by tonics and a change of climate; that of the gouty woman is cured by agencies directed against the diathesis, etc. Of these states of the endometrium, I will not treat. The neglect to draw clearly the distinction between those changes in the uterus which are hematogenous and those produced by germs has led to much confusion among gynecologists regarding the various methods of treatment. It is in the first class that applications to the endometrium are permissible. I can but repeat the caution I gave above, for it is exceedingly easy by means of a filthy sound or dilator to change one type without complications into that other virulent form which is the starting point of processes of the most destructive kind.

Forms of Endometritis.

Septic,
Gonorrheal,   } Of the Cervix;
Tubercular,   Of the Corpus.

ACUTE AND CHRONIC SEPTIC ENDOCERVICITIS.

The dense, firm tissue of the cervix has normally great resistant power against all the pyogenic cocci except the gonococcus. Lined as it is by a membrane well supplied with racemose glands, it presents the characteristics of other structures so formed. In the mature woman septic endocervicitis does not often exist alone, but in young women it is very common (Fig. 1.) Most often it is found to be co-existent with a more or less severe type of corporal endometritis. Where the gonococcus is the cause of the inflammation, endocervicitis very frequently exists alone.

Symptoms.—In the acute stage, beyond a sense of weight in the pelvis, little discomfort is felt. There is a profuse purulent discharge from the cervix, tenacious and hard to remove. But in some cases no discharge is noticed other than one normal to the parts, the pathogenic germs being quiescent. Upon examination we find the cervix congested and in very acute stages, bleeding upon the slightest touch. The follicles project from the surface as red papillæ and give to the cervix the appearance of being " eroded " or " ulcerated." (Figs. 2, 3). In more chronic cases the Nabothian follicles become affected; certain of the typical glandular follicles become closed, and they too form cysts. This process of inflammation may be so general and extensive that the

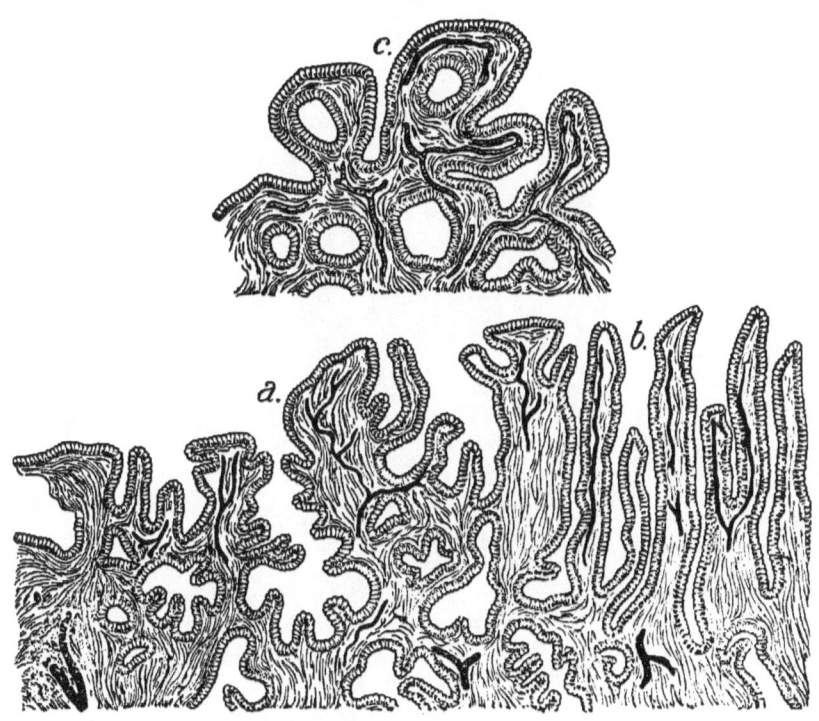

FIG. 2.—*a*, *b*, simple papillary erosion; *c*, follicular, slightly enlarged (Pozzi).

ENDOMETRITIS.

cervix becomes riddled with cysts, constituting *cystic degeneration*. In such cases, in addition to a possible folliculitis, we have presented the appearance of slight rounded elevations projecting upon the vaginal face of the cervix and covered by epithelium. Upon pricking

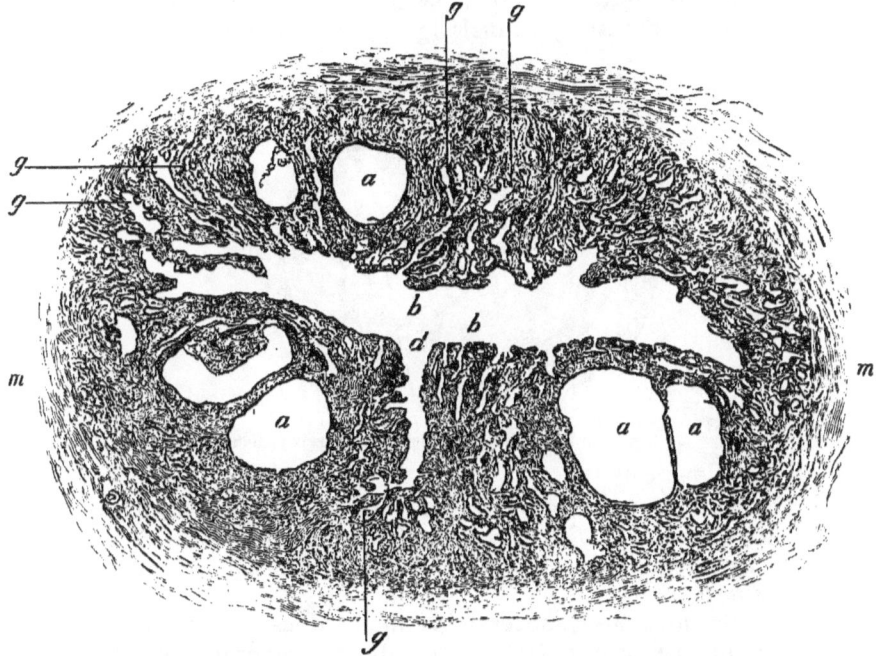

FIG. 1.—Transverse section through the upper part of the cervix, showing the entire mucous membrane. The central cavity is the cervical canal: b, b, internal surface of mucous membrane, presenting small folds, superficial glandular depressions, and large incisions of the arbor vitæ (d); g, g, deep glands; a, a, ovules of Naboth; m, m, muscular tissue of the uterine wall (Cornil).

one, either a clear, glairy fluid escapes, or else a drop of pus is squeezed out, leaving a depression. The condition obtains not only upon the surface, but also throughout the entire cervical structure. When the mouths of the glands within the cervix become closed, they may continue to secrete, and the enlargement becomes more or less pedunculated, forming a *polypus* (Fig. 4). Even in cases presenting no symptom other than enlargement, a positive diagnosis of the cause can not be determined without the microscope. In collecting the secretion, a curette should be used and not a swab, as pressure upon the glands is necessary to dislodge their contents. Where

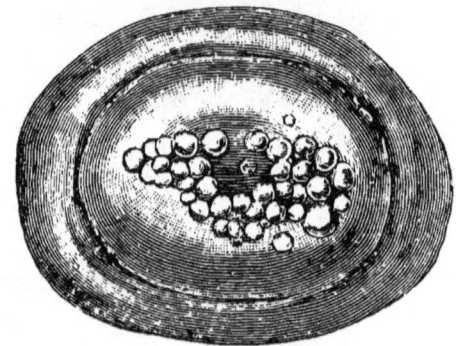

FIG. 3.—Simple follicular cysts of the cervix (Auvard).

cystic degeneration is present, emotional disturbances are common. *The presence of any purulent discharge calls for not only a digital examination, but an ocular one as well.*

Diagnosis.—Endocervicitis alone does not produce the intense pain, fever, and the uterine tenderness which accompany purulent endometritis. The erosions of endocervicitis can scarcely be mistaken for either the ulcerative or nodular cancer of the cervix; but a badly cystic cervix will sometimes closely resemble nodular cancer. The patient's age may still further render the case suspicious. If cancer exist, a pair of blunt bullet forceps made to grasp the questionable cervix will tear out easily and produce profuse bleeding. In cystic degeneration there is greater strength in the tissues, and pulling the forceps through such a cervix is very difficult and will

ENDOMETRITIS.

evacuate a number of cysts. It is commonly easy to push a Simpson's sound into a cancer nodule. Cystic degeneration is apt to be general throughout the cervix, whereas nodular cancer is at first—the only period at which it is possible to mistake it for any other condition

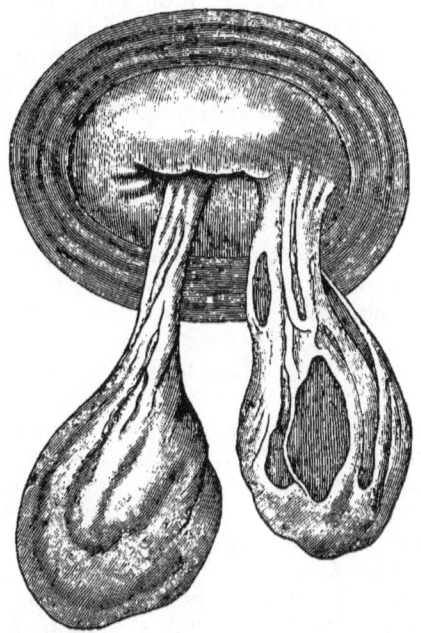

FIG. 4.—Mucous polypi from the interior of the cervix and upon the surface (Pozzi).

—limited to one portion of the cervix, and even later on in the disease one nodule is prominent over the others. The nodules of cancer are larger than the cysts and have more injected and more vascular edges. It must not be forgotten that both cancer and cystic degeneration may exist in the same subject. Should the diagnosis be still further in doubt, a piece should be amputated and submitted to a pathologist for examination.

Treatment.—Endocervicitis should be vigorously treated. The best application is tincture of iodin, a powerful diffusible antiseptic and astringent. A silver applicator is wrapped with cotton and dipped in the iodin. It is then passed *to* but not through the internal os. The cervical speculum should not be used. The cervix, if markedly congested, should be punctured with a scalpel in a half dozen places to produce bleeding, and this should be promoted by hot vaginal douches of ¼ of 1 per cent. lysol every four hours, continued for one day. The iodin is to be applied every other day except in gonorrhea, and then once each day. Usually, in a few days this treatment will subdue all symptoms of inflammation. If the cervix be the seat of polypi, or is hypertrophied or markedly cystic, it should be amputated after all acute symptoms have subsided. During treatment coition is prohibited.

The latency of gonorrheal and septic endocervicitis and the fact that either form of infection may exist without producing pathological discharges must be constantly before the physician when he wishes to use the sound, or to operate upon the cervix. In any case giving the history of a previous purulent discharge or any form of pelvic inflammation, the introduction of the sound through the unsterilized cervical canal is positively contra-indicated. The application of tincture of iodin to the cervix will superficially sterilize it for the use of the sound. If *cystic degeneration* is present the cysts should be pricked and the pits touched with tr. of iodin. If *cervical polypi* are present they may be removed without narcosis by using cocain. An applicator wrapped with cotton is dipped in 10 per cent. cocain solution and passed into the cervix. It is left there five minutes. The polypus is then seized by Luer's forceps and twisted off. The bleeding is usually trivial. If of moment, the cervix is steadied by bullet forceps and iodin painted over the oozing surface, after which the cervix and vagina are packed for twenty-four hours with iodoform gauze. These polypi, although not of cancerous nature, are apt to recur, and it is therefore advisable to amputate the cervix by Shroeder's method if more than one polypus is present.

ACUTE GONORRHEAL ENDOCERVICITIS.

The gonococci lie upon the surface of the membrane and deep in the glands. Of all acute forms of endocervicitis, this is the most common.

Symptoms.—These are the same as those of septic endocervicitis. With gonorrheal infection there is apt to be more erosion and the production of pus is greater.

The diagnosis is based upon finding the gonococcus in the discharges and upon the presence of symptoms of gonorrhea in other parts of the genital tract or urethra.

Treatment.—Strong solutions of silver nitrate ($\mathfrak{z}$j to f$\mathfrak{z}$j) may be applied daily so long as acute symptoms exist, but the tincture of iodin and local blood-letting have given me such excellent results that I usually employ them. A slender applicator is wrapped with cotton and dipped in the solution. It is then inserted into the cervix up to the os internum. This application is made every day or two until the discharge of pus ceases. The vagina should be douched with bichlorid, 1 : 10,000, every three hours. The complications resulting from an extension of the infection are gonorrheal endometritis and salpingitis. The cervix infected by the gonococcus should never be irrigated lest the cocci be washed higher into the more important uterine cavity.

CHRONIC AND LATENT GONORRHEAL ENDOCERVICITIS.

This condition is more common. A woman may have the gonococcus infect her cervix and pass through a mild attack of "yellow leucorrhea" without disagreeable subjective symptoms. Absolutely all discharge may cease, and yet she be in a condition to infect others and to become acutely inflamed if the cervix be subjected to traumatism. I examined one hundred clinic cases showing no discharge of pus from the cervix. The gonococcus was found in the cervix in twenty-two. This fact of latency explains those accidents which follow the rough use of the sound or operations upon the cervix. The cervix must be scraped with a small curette in securing its discharge, for the cocci lie deep in the glands.

The symptoms and treatment are the same as for septic endocervicitis. In fact the clinical picture of both conditions is much the same and bacteriological examination will alone differentiate them.

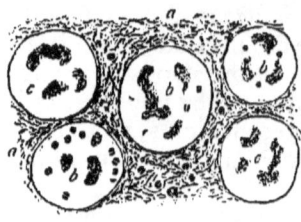

FIG. 5.—Gonococci in the secretion from the urethra in fresh gonorrhea: *a*, mucus with separate cocci and diplococci; *b*, pus cells with diplococci; *c*, pus cells without diplococci; × 700 diam.

After repeated attacks of cervical inflammation the tissues about the cervix become infiltrated by connective tissue elements, and as a result the cervix becomes more or less fixed at the vaginal vault. This pericervical thickening and contraction is more marked in the condition described as *genital sclerosis* (Fig. 5).

SEPTIC ENDOMETRITIS.

ACUTE SEPTIC ENDOMETRITIS.

Septic inflammation of the lining membrane of the body of the uterus is not often found to exist in cases which have not recently been pregnant or which do not have neoplasms, as polypi, within the uterus; but the condition existing as a chronic state may at any time be converted into an acute one by any traumatism inflicted upon the uterus. The causative germs are staphylococci and streptococci (Figs. 6, 7 and 8).

Symptoms.—There may be a chill, but this is not usual. From twelve hours to three days after the infection the patient begins to have a sense of fulness in the pelvis. In a few hours this increases to a positive pain,

ENDOMETRITIS.

and in aggravated cases spasmodic contractions of the uterine muscle follow ("womb cramps"). With the first

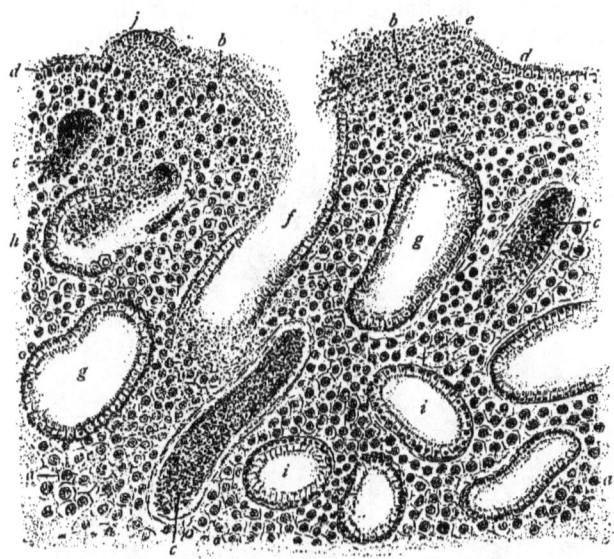

FIG. 6.—Normal mucous membrane of the uterus during menstruation. A preparation of the layer removed by curette during menstruation. The figure reproduces the upper third of the membrane. There are small extravasations here and there; in the deeper parts is almost normal interglandular tissue, the glands being somewhat more sinuous than usual. The blood-vessels are full; the upper layers are partly normal, partly infiltrated with blood cells; the epithelium, for the most part preserved, is here and there raised from its seat and covered with altered blood cells; hemorrhage into the glands in places; no appearance anywhere of the fatty degeneration described by certain authors (Williams, Kundrat, Engelmann). It is very likely that sometimes during menstruation part of the mucous membrane is destroyed (Leopold, Wyder), and that there is no such desquamation at other times (Moericke). This figure shows that the different changes may be simultaneous, and that there is great diversity in the process : *a*, Normal uterine tissue formed of numerous rounded embryonal cells ; *b*, same, infiltrated with blood corpuscles to a considerable depth ; *c*, dilated vessels, full of blood ; *d*, intact uterine mucosa ; *e*, place where it has become detached ; *f*, longitudinal section of a gland ; the epithelium near its mouth has disappeared ; *g*, dilated glands ; *h*, gland whose lining epithelium has become detached ; *i*, normal deep glands ; *j*, mucous membrane raised by infiltration of blood (Pozzi).

symptom there takes place a slight rise in the temperature, and the evening temperature is usually one degree higher than the morning. Or, there may be no fever. The subjective symptoms rapidly increase until within a few hours the discharge begins. At first this is mucopurulent, but generally by the end of the first day it has become distinctly purulent and may be tinged with blood. The woman takes to bed, so great is the suffering, and

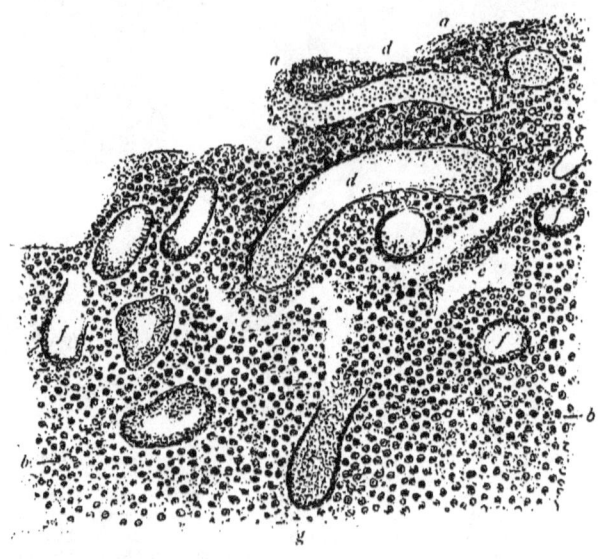

FIG. 7.—Acute endometritis. Slightly enlarged view of entire mucosa: *a*, Superficial layer formed of more or less altered tissue, infiltrated with coagulated blood; *b*, round-celled embryonal tissue; *c*, zone in which these cells are especially numerous; *d*, large dilated and varicose vessels gorged with blood; *e*, lymph spaces; *f*, transverse sections of glands; *g*, glandular cul-de-sac (Pozzi).

soon after the onset the pain has become general over the pelvis. If the infection occurs at the menstrual time, the flow is increased considerably and the blood is apt to clot.

Upon *examination* the suprapubic region is sensitive. The finger introduced into the vagina passes without evidence of suffering until the cervix is touched, when the patient will jump and utter an exclamation. If

ENDOMETRITIS.

bimanual examination is possible, the entire uterus will be found exceedingly sensitive and more or less fixed in the pelvis. This fixity is not due to peritonitis or tubal disease in all cases, but to a tonic contraction in the ligaments supporting the organ. Examining with the speculum, a rope of mucopus is seen hanging from the cervix.

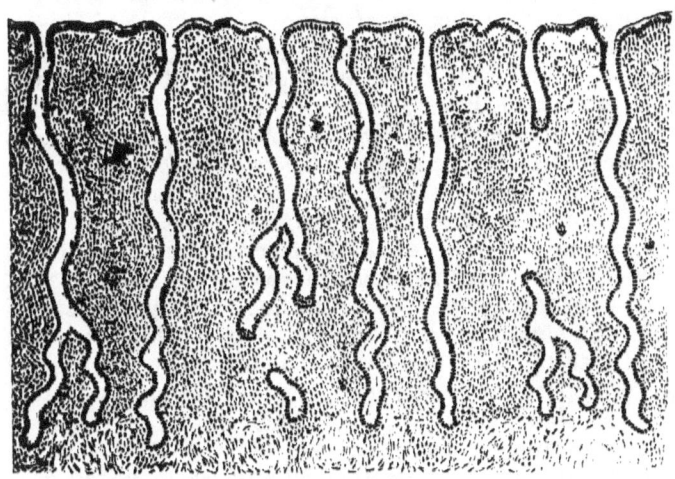

FIG. 8.—Microscopic section of the normal endometrium, showing the utricular glands extending into the muscular tissue (Beyea).

The cervix is deeply congested, and in cases due to infecting plastic operations upon the cervix the tissues may be in a sloughing condition. Wherever there has been a break in the cervical or vaginal tissues there will be found a patch of gray false membrane. This is only occasionally the case in staphylococcus infection, but almost invariably in streptococcus inoculation. The cervix may bleed upon the slightest touch; it is soft and the os more open than normal. After existing for a few days, the case either presents the symptoms of some complication or the acute symptoms subside. In the latter event, after three days the woman feels better, the discharge diminishes, and within a week she is able to be up. This happy result is not often seen in untreated cases, and in those cases which do recover without complication, the inflammation usually persists as a chronic process.

Differential Diagnosis.—There is usually a history of traumatism, such as an operation or invasion of the inside of the uterus or of abortion or labor. This is lacking in *gonorrhea*. Again, in gonorrhea there is to be found usually some other manifestation of that disease, as urethritis or Bartholinitis, or vulvitis. *Endocervicitis* does not present the grave symptoms which are due to endometritis. Acute *tubal and ovarian inflammation* and peritonitis give signs in the peri-uterine tissues which are not found with uncomplicated endometritis.

Sequelæ.—The most common sequela of a neglected septic endometritis in a nulliparous uterus is salpingitis. The peritoneum may become involved, and from this point the infection may spread so as to implicate the ovaries and general pelvic peritoneum. The uterus may be retroposed or become so, and will become adherent in its displaced position.

Treatment.—NON-OPERATIVE.—If the condition follows a septic plastic operation, all sutures should at once be removed and the wound painted with tincture of iodin. No vaginal dressing should be applied, but the freest possible exit to the pus afforded. Warm vaginal douches of one-half of one per cent. lysol or three per cent. boric acid are to be used every three hours. Attempts should always be made to wash out the uterus. The ease with which this may be done is dependent upon the state of the cervical canal. The patient is placed in Sims' position and the perineum drawn back. The operator and his material are sterilized (see sterilization). The cervix is sterilized by means of an application of iodin (see endocervicitis), steadied by a pair of blunt bullet forceps, and the direction of the uterine canal found by a sound. A Fritsch-Bozeman catheter to suit the size of the cervical canal is then introduced up to the fundus (Fig. 9). No vio-

FIG. 9.—Irrigating the uterus through a double-current catheter.

ENDOMETRITIS.

lence is to be used. The irrigator (see sterilization) should be five feet above the patient. At first a quart of Thiersch solution is allowed to flow through the catheter, to be followed by at least four quarts of a three per cent. solution of boric acid crystals. The treatment varies according to the progress of the case. Often one such washing will suffice to subdue acute symptoms; but, if after waiting twelve hours, the patient is not markedly better, the washing is to be repeated. The irrigations are to be made once or twice daily, the physician being governed by the amount of discharge and the symptoms. An ice-bag should be worn over the pubes continuously until convalescence begins. The sense of weight and even the inflammation are materially lessened by local blood-letting. This is done by superficial stabs into the cervix with a bistoury.

After the acute symptoms subside, the case is to be treated as are cases of chronic endometritis. The pain is sometimes unbearable. In vigorous individuals phenacetin, grs. v, with codein, gr. ss., may be administered and repeated in two hours if needed. Or a rectal suppository of extract of opium and ext. belladonna, each gr. ss., may be given. But in administering these drugs the sympt ms are so masked that the extension of the disease may not be appreciated, and it is therefore advisable to avoid them. It is not necessary to purge these patients; merely normal stools are all that are required. The rectum must be kept empty.

If after two days' treatment the local and general symptoms do not improve, an extension to the adnexa or peritoneum is to be suspected. During the treatment, light vaginal tampons of iodoform gauze may be used instead of the douches, where the uterus is subjected to jarring by vomiting.

SURGICAL TREATMENT.—In view of the possible extension of the infection to the peritoneum and adnexa, it is important to check the disease at once. This can be done with certainty by a properly performed curettage (see curettage). The responsibility resting upon the attendant is so great that he should in all cases place himself clearly on record with his patient and compel her to assume responsibility for any complication which may follow a neglect to clean out an infected uterus. If the curettage has been improperly performed or done too late to check an extension to the tubes and peritoneum, it will be necessary to open the cul-de-sac and treat the adnexa (see cul-de-sac operation).

PUERPERAL INFECTION.

This is an infection occurring during the first four weeks after delivery. Infection ensuing after that time is not puerperal infection, but is merely endometritis in a large uterus; and treatment applied to an infected uterus after the puerperal month is not the treatment of a puerperal uterus, although the lesions may be the result of a puerperal infection. Therefore, curettage or hysterectomy done some six weeks after delivery is not to be considered as having been applied for puerperal fever. The condition of the tissues one week and six weeks after labor are so different that the lesions produced are different, and the dangers from a hysterectomy at the two intervals are about as fifty per cent. to five per cent. Infection after *abortion* is similar to that after labor. But the smaller uterus with its less active lymphatics and vessels, when infected, produces less septicæmia. The gravity of the symptoms is usually in direct ratio to the period of gestation. The infections after early abortions, as at the fourth week, take on the type of endometritis. The later abortions assume the characteristics of infection at full term. The line cannot be sharply drawn between those cases to be classed purely as abortions and those which shall be called puerperal.

It is eminently proper that I describe in a separate chapter the forms of infection occurring during the puerperium. More especially am I prompted to do this because the method of treatment I employ is somewhat different from that of most surgeons and because these cases commonly fall into the hands of the general practitioner.

The puerperal uterus may be the seat of an invasion by any one of the pyogenic cocci and various bacilli. The lesser degrees of infection are caused by staphylococci and the more virulent by streptococci. These produce *septic endometritis*. Certain saprophytes when introduced into the puerperal uterus produce *putrid endometritis*.

The activity of these microbes results in the production of toxins which alter the chemical processes in the body and may cause death.

The infection starts in the endometrium and may be general over the whole inside of the organ or limited to the placental site.

If *putrid* infection occurs there will be found over a certain area of the endometrium a patch of slough in which the saprophytes are situated, more or less mixed with other germs. Beneath this patch there is arranged in the endometrium an aggregation of white blood corpuscles whose presence tends to prevent an extension of the saprophytes into the deeper parts of the organ.

In *septic* infection the cocci lie either upon the surface of the endometrium, within it, or may even penetrate into the peri uterine structures.

Putrid infection results in a sapræmia, while septic infection causes a septicæmia.

Putrid infection, pure and unmixed with septic germs, is a superficial affection and does not destroy life. But inasmuch as the necrotic area is so likely to become the point of entry of septic germs, the appearance of sapræmia is often the forerunner of a septic process.

Septic Puerperal Endometritis.—The cocci enter the walls of the uterus through the lymphatics or the veins. As they proceed we have perhaps the veins filled with infected blood clots, *thrombo-phlebitis;* or there may be a pelvic, then a general, *lymphangitis* (Fig. 10). The peritoneum lying near the infected spots throws out a quantity of serum and lymph and a peritonitis results. Immediately

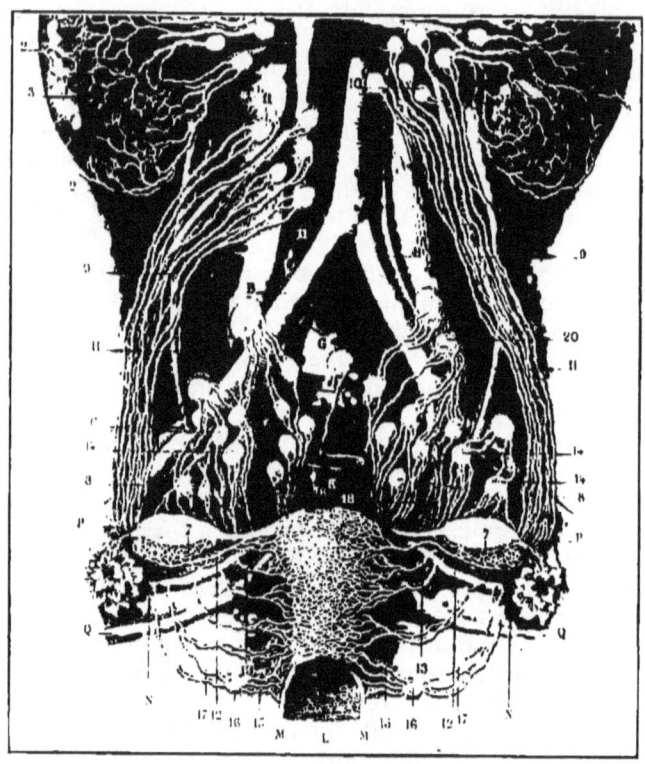

FIG. 10.—Lymphatics of the pelvic viscera and abdomen: A, Aorta; B, B, iliac arteries; C, C, the bifurcation and two branches of the iliac arteries; D, vena cava; E, left renal vein; F, right renal vein; G, iliac veins; H, H, ureters, I, rectum; K, uterus; L, cervix; M, M, vaginal walls; N, N, Fallopian tubes; P, P, ovaries; Q, Q, round ligaments; 1, Deep lymphatic vessels of the right kidney, and ganglia into which they empty; 2, 2, 2, 2, superficial lymphatic vessels; 3, 3, 3, 3, the same; 4, two ganglia that receive these superficial vessels; 7, 7, subovarian plexus of lymphatics; 8, 8, ducts leading from this plexus; 9, 9, the same; 10, 10, 11, 11, glands receiving these ducts; 12, 12, 12, 12, lymphatic ducts, originating in the fundus uteri, and terminating in the same glands as the ovarian ducts; 13, 13, ducts from the anterior surface and sides of the uterus; 14, 14, glands into which they empty; 15, 15, ducts originating in cervix and upper part of vagina; 16, 16, glands into which they empty; 17, 17, efferent vessels of these glands; 18, 18, lymphatic ducts from posterior surface of the uterus and glands into which they empty; 19, lumbar gland (exceptional); 20, gland into which occasionally a duct from lower uterine segment empties (Sappey).

ENDOMETRITIS.

adjacent to the infected vessels suppuration may occur, and this may take place primarily in the folds of the broad ligament or in the ovary, and subsequently the fallopian tube and pelvic peritoneum may be the seat of a suppurative process (Fig. 11).

In some cases the gross lesions are all local, but occasionally as in cases reported by the author as early as 1886, and again in 1895, specimens have been shown in fatal cases where there were no gross lesions observable in the pelvis, but only at some distant point, such as the diaphragmatic peritoneum, the pleura or the heart membranes.

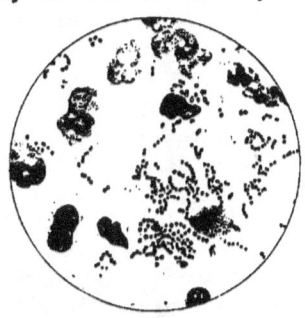

FIG. 11.—Streptococcus pyogenes in pus (× 1000) (Fränkel and Pfeiffer).

The more virulent and rapid the infection the less are the local manifestations of the disease in cases which die. There is in some instances produced a stasis in the infected uterus, and before it is possible for masses of lymph to be effused or pus to be produced the patient is dead. A fatal result of course more speedily ensues in the thrombo-phlebitic type. If the patient recovers through the action of natural processes only, it will be with damaged pelvic organs. The pelvic lymphangitis results in the production of connective tissue in the broad ligaments, which bind and fix the uterus. The effusion of lymph upon the pelvic peritoneum results in the occlusion of the tubes and all the various forms of adhesions to be found in the pelvis. There may also be found pus foci in the broad ligament, the ovary, or even, as a secondary complication, in the tube. The organs higher in the abdomen may be bound to each other, and there may be adhesions in the diaphragmatic pleura. As a result of endocarditis the valves become distorted and permanent heart lesions are found. Throughout the body, along the lymphatics of the supra-clavicular spaces, in those of the groins and at other points buboes may form and result in disfiguring scars. A thrombophlebitis in a leg may permanently interfere with its circulation, and articular inflammations result in stiff joints. Finally, any one or all of the results of a profound degree of general septicæmia may be noted. The graver sequelæ are endocarditis, pneumonitis and nephritis. Naturally women who have suffered from septicæmia are particularly prone to contract phthisis.

It is not the province of this book to enter more elaborately into a description of the methods of contagion, etc., for these points are elaborated in many works on obstetrics. I have stated enough to show the importance of early treatment, not only that life may be saved—we have passed beyond that—but that the ravages of sepsis in the system may be prevented. We have ceased to be guided solely by the mortality, and are attacking the morbidity statistics (Fig. 12).

Post partum women may have a rise in temperature from general causes as well as from intra-uterine infection. Before proceeding to treat the uterus, the responsibility for the rise in temperature must be placed there. It is taken for granted that so soon as the pyrexia has been noticed, the attending physician has eliminated the element of intestinal toxicosis by washing out the colon with normal salt solution, and likewise has proven the case non-malarial by the administration of a large dose of quinin. It is an undoubted fact that seventy-five out of every hundred women who have a rise in temperature after labor, will reach a normal state in a few days, if left alone. But among the other twenty-five will be found a few who will perish in a few days if not treated, and the others of that number will be invalids for life. It is most desirable that every physician in America have the ability to make cultures from the inside of the uterus, and in this way be able to differentiate the various causes of puerperal fever. Still, few of the expert bacteriologists agree upon one plan of collecting these discharges, and they are still debating about it. The

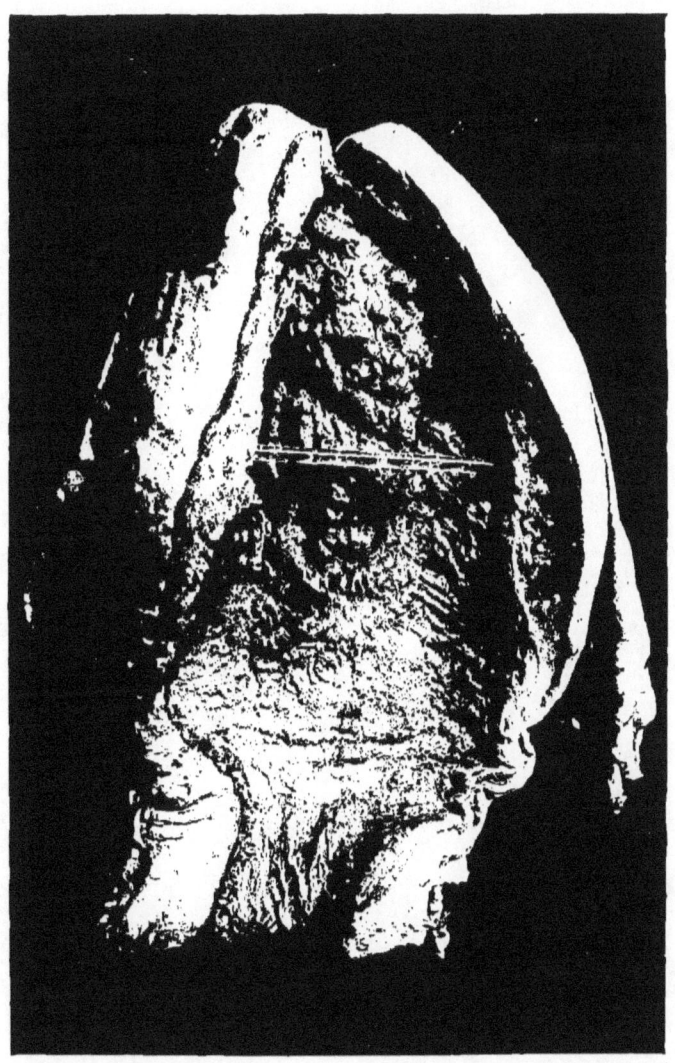

FIG. 12.—Specimen from a patient who died septic, showing the material that would be found to be removed by the curette or the finger on the "roughened placental site." "Clots in the uterine sinuses" (Army Medical Museum, Washington, D. C., No. 10,619). ("An American Text-Book of Obstetrics.")

ENDOMETRITIS.

fact remains, anyway, that the great body of the profession prefers to be governed by clinical symptoms rather than bacteriological examinations when made by any except the most skilled. Therefore, the family physician who attends most of these cases of puerperal fever must either waste important time while determining that a given case is virulent streptococcus infection or not, or else he must apply that method of treatment which will do good in all cases and harm in none: At least, any sensible man will, though a theorist may not.

Having flushed out the bowel, cinchonized the patient and having excluded all other causes for the fever, if the patient evidently suffers from an infection starting from the uterus, I proceed as follows:

Treatment.—The patient's vagina should be thoroughly cleansed by a prolonged douching with bichloride of mercury solution, 1 : 10,000. The operator prepares his hands and material as described in the article upon sterilization. All of this is done before an examination is made. Inasmuch as putrid infection remains superficial and sepsis at one stage is also only upon the surface of the endometrium, I advocate intra-uterine douches. The time element is important. If seen within twelve hours of the onset of the first evidences of infection, before the germs have penetrated the deeper layers of the decidua, intra-uterine irrigations will cure. But if the patient is seen after the germs have passed into the decidua, some antiseptic must be employed which will be more rapidly absorbed than they can proceed, as iodin in the form of iodoform. If they have passed still deeper, curettage is needed.

Should the invasion have passed outside the uterus, curettage and the cul-de-sac incision with proper dressings applied to the pelvis will cure nearly every case. The excepted cases are those in which purulent peritonitis has set in, cases of thrombo-phlebitis, and those with some mortal complication, as endocarditis. Rarely will hysterectomy be indicated, so rarely in fact that I advise against it except under the circumstances which surround one in a perfectly appointed hospital.

Puerperal infection is wound infection and should be treated as such.

IRRIGATION OF THE UTERUS.—The vulva should be shaved and the patient placed in the lithotomy posture on the table, with a suitable arrangement made for catching the irrigating fluids. A sheet tightly rolled and shaped like a horseshoe may be laid under the buttocks, and a piece of rubber cloth over this will protect the clothing and floor. The vagina is irrigated with four quarts of bichloride solution 1 : 10,000. The perineum is drawn down by either a speculum or the fingers and the anterior lip of the cervix is seized with my blunt bullet forceps (see Fig. 42) and drawn down. The largest return-flow irrigating tube (see No. 4, Fig. 49) is then passed to the fundus. In doing this the instrument is not allowed to touch any part of the patient until the cervix is reached, and the utmost gentleness is employed to avoid bruising the endometrium. The tube of the fountain syringe is then attached and the current turned on. I first allow one quart of Thiersch solution to pass into the uterus, to be followed by six or eight quarts of boric acid or salt solution (see Fig. 9).

Frequently this one washing will suffice to control all symptoms. The operator remains in the house for an hour and repeats the procedure. The rectal temperature is taken every half hour. If in four hours the temperature is not normal, the uterus is again irrigated, and is packed full of iodoform gauze, 10 per cent. strength. This packing is made in the following manner: The strips of gauze are four inches wide. While steadying the uterus with the heavy forceps, the operator seizes the end of the gauze in Hunter's forceps, and passes it to the fundus. He continues to insert the strip until the uterus is filled, the end of the gauze extending into the vagina. The vagina is then packed with iodoform gauze. The iodoform becomes rapidly disintegrated in the presence of the uterine discharges and free iodin is detected in the urine in three hours.

If in twenty-four hours after this, the temperature is

not normal, the operator proceeds to curette the uterus and open the cul-de-sac; for the infection being no longer superficial, it is impossible to say how far it has extended; presumably it has passed to the peri-uterine lymphatics, and it will be necessary to treat these by sterilizing dressings.

But if the temperature is normal in a day, the packing is allowed to remain in the uterus for forty-eight hours and is then withdrawn. It is not renewed. It is almost needless to say that the presence of retained membranes or placenta requires their careful renewal before instituting this treatment. I must advise against digital exploration of the uterine cavity, unless it is followed by irrigation and gauze packing. It is most difficult to disinfect the finger, whereas instruments are easily cleansed by boiling. Whenever I suspect the presence of placental tufts or membranes their removal is to be made under chloroform by Luer's forceps or Mundé's curette.

And if retained placenta is accompanied by symptoms of sepsis, its removal is to be accompanied by curettage of the uterus; for it is presumed that under the circumstances the infection also invades portions of the endometrium other than the placental site.

CURETTAGE.—The uterus is drawn down as before and washed out with several quarts of bichloride 1 : 10,000. The antiseptic is not contraindicated here, for it is intended to remove the decidua and no slough will follow its use. Taking the largest sized curette, the operator introduces it into the uterus and systematically curettes the entire surface. It will not be necessary to use force as the tissue is soft. The bleeding is pretty free, but is to be ignored. After curetting the organ, it is again washed out with the large irrigator, which has been resterilized while the curettage was proceeding. At this second washing I employ boiled salt solution at a temperature of about 115° F. The uterus is then packed full of iodoform gauze, 10 per cent. strength. The size of the cervical canal will largely govern the width of the strip to be used. As the patient is under chloroform a large strip can be inserted, usually nine inches in width. The organ is to be tightly packed, and the vagina also filled. In two days these dressings are removed, and a second packing made, but not so firmly. The cervix will be found widely open. This second dressing is taken out in two days more and the patient is put on ergot and quinia.

If this treatment has not subdued the symptoms in forty-eight hours, the cul-de-sac must be opened.

THE CUL-DE-SAC OPERATION.—Whenever I curette the uterus for *sepsis* I immediately open the cul-de-sac. In cases of putrid infection and where retained placenta is sought curetting alone will suffice. But in the presence of sepsis which has not yielded to the non-operative procedures, I have never been able to determine how deeply in the tissues the infection has extended. And as the cul-de-sac operation is devoid of danger, and my anxiety is great lest the infection run away from my reach, I always open the cul-de-sac whenever I curette for sepsis, *i. e.*, when milder procedures have failed.

The patient is under chloroform. The vagina is prepared as described elsewhere; and if there could be degrees in cleanliness I would urge the highest here.

After curetting the uterus it is packed with gauze. Selecting the fold just behind the cervix, it is picked up with forceps and cut through for a space of a half-inch. The cut extends through the mucous membrane only. The finger is then shoved into the pelvis. About two minutes are consumed in this. Upon withdrawing the finger quite a quantity of serum escapes. The operator carefully notices the character of the fluid which escapes. He passes his finger back and forward behind the broad ligament and gently palpates the adnexa. If the organs are found matted together by lymph, they are liberated with the finger. I seek to make this digital exploration and separation of adhesions of the broadest kind, my object being to open the lymph spaces not only that they may discharge into the dressing which I am about to apply, but that the antiseptic may be readily absorbed.

Wherever there is an effusion of lymph there is a contest between germs and the tissues. Into this combat I wish to enter.

Having examined the pelvis and separated all adherent organs, I insert two fingers into the cul-de-sac and stretch the opening to a level with the sides of the cervix. The pelvis is then filled with strips of iodoform gauze (see Fig. 38). These pieces are made by taking a strip of gauze a yard long and four inches wide, and folding it until the shape represented. The strips extend up to the level of the fallopian tubes. They are inserted as described in the article on hysterectomy. The vagina is packed. If the patient's pulse runs under the ether to 140, I deem it the indication of a profound degree of septicæmia. It is then my duty to increase the action of the kidneys as well as stimulate the heart. I therefore introduce into a vein from one to two quarts of salt solution (see salt solution). This I also do whenever I find that the lymph in the pelvis is breaking down into pus, and whenever there are grave complications as pneumonia and nephritis. It is a question in my mind whether it be not advisable to employ this in all cases of streptococcus infection and at present such is my practice, but I am not prepared to advise its general use.

Within a day after this operation the bed will be found soaked with the muddy, toxin-laden serum from the pelvis. The amount must be pints in quantity in some cases, and the patient will feel the loss of so much fluid unless provision be made to supply it. I therefore inject into the bowel eight ounces of tepid salt solution every three hours for several days. The patient is stimulated by hypodermics of strychnia, but I do not believe in excessive doses. I usually give gr. 1:50, q. 4. h. A few hours after the operation I begin giving fluids by the mouth. If stimulants are needed I give either an ounce of champagne every hour or a teaspoonful of brandy in an ounce of water. If alcoholics are not necessary, I give hourly an ounce of cold water to which has been added five drops of lemon juice. The urine is drawn every three hours and is measured. It is tested for albumen and iodin. I begin to administer liquid food after eighteen hours, beginning with a little hot chicken-broth or squeezed juice of broiled steak. In three days I remove the vaginal packing and the uterine gauze. The vaginal packing is renewed, but the uterine is not unless the septic symptoms persist. The cul-de-sac dressing is taken out in a week and is replaced by fresh gauze of 5 per cent. strength.

These dressing are repeated every four to seven days until the opening closes. After that I apply ichthyol tampons to promote involution. In conjunction with Dr. Jeffries, bacteriologist to the Polyclinic, I have instituted a series of experiments in the influence of these dressings upon the streptococci usually found in these cases. I have in fifteen cases placed the uterine scrapings in a sterile tube and the fluid from the cul-de-sac in another. Whenever streptococci have been found, even though pus was free in the peritoneal cavity in one case, and sacculated in several others, we have never failed to find that the dressings absolutely sterilized the operation field. This is usually accomplished by the third dressing, but in one case not before the fifth had been used. And in many cases where no bacteriologic examinations were made the treatment was equally successful.

So far as results are concerned, I have not lost a patient so treated. The success of this operation in a class of cases formerly thought worthy of hysterectomy convinces me that the latter operation is unnecessary. And inasmuch as the operation is exceedingly simple and requires no elaborate equipment of nurses, assistants, and material, I shall expect it to become of general adoption by those who have the surgical conception of the treatment of this dreadful disease.

The importance of the proper treatment of this form of infection is well shown by the report of the Registrar General for England and Wales for the year 1895. Out of every thousand deliveries two women died from some form of infection. The morbidity in those that recovered must have been appalling.

ACUTE GONORRHEAL ENDOMETRITIS.

The sole caus-tive germ is the gonococcus, but the infection is often a mixed one, other cocci besides the gonococcus being present. Occasionally a woman will have gonorrheal vulvitis, vaginitis, or endocervicitis for some time without an extension to the endometrium; but prolonged exposure, menstruation, overindulgence in coition, and any operation upon the uterus, or even the passage of the sound, will cause a sudden attack of gonorrheal endometritis.

Symptoms.—These cannot be better brought out than by describing a case. A girl of twenty-two married two weeks before an expected period, a man with gleet. Five days afterward she began to have painful and frequent urination, and in a few days more a vulvovaginal swelling appeared upon the right side. She thought these symptoms due to frequent connections, and bore her distress with good grace. There was a profuse purulent discharge from the vulva, and she was compelled to bathe frequently. The menstruation appeared on time, and was normal up to the third day. She then had sharp, lancinating pains in the uterus accompanied by the most severe cramps. The menstrual flow increased in amount, and three days after the onset of the severe symptoms she had a most irritating and profuse yellow discharge. She took to bed. The entire pelvis became so tender that she could not bear to be touched. The pus was mixed with blood. The rectal temperature when I saw her on the tenth day was 101.8° F., and the pulse was 106. Upon examination, I found a vulvovaginal abscess discharging. The uterus was exquisitely sensitive, and the uterine spasms continued after the menstrual flow ceased. The cervix was livid in hue, and markedly eroded. It seemed to be entirely devoid of epithelial covering, and from the os hung a rope of tenacious yellow discharge. The microscope showed the gonococcus.

Such is a picture of an average case. The uterine

infection may take place at any time, but it is most apt to occur during menstruation, when the uterine epithelium is exfoliated. Several days elapse, as a rule, between the onset and the appearance of pus. The pus is produced in great quantities, as much as several ounces in twenty-four hours. It is tinged with blood in nearly all cases, so deep is the congestion. The body temperature is elevated, but rarely goes to 103° F. The pulse is accelerated up to 110 in the worst cases. The symptoms subside slowly, but the purulent discharge continues for some time, and, as a rule, the case becomes chronic. Rarely does a cure occur without treatment, and complications are very common. The uterus in the acute stage is enlarged; but after repeated attacks it may become little more than a mass of fibrous tissue, being hard and small. Such uteri we find in old prostitutes. Women with acute gonorrheal endometritis take to bed. The suffering is continuous, and the pain is marked by sharp exacerbations—" uterine colic." The course of an attack persists through one or two weeks and results in either a chronic condition or some grave complication.

Diagnosis.—There will usually be found other evidences of gonorrhea, such as urethritis or vulvitis. A woman previously well suddenly attacked with acute endometritis a few days after connection probably has gonorrhea. By far the greater number of such cases of acute endometritis which do not occur after abortion or labor are due to gonorrhea. Indeed, I am warranted in saying that gonorrhea is a disease of the non-pregnant uterus, while sepsis is most frequently found to follow conception.

The temperature, the great pain in the uterus, the profuse purulent discharge, the excoriations produced in the cervix and vagina by the pus, the presence of other evidences of clap will render the diagnosis clear. In all cases the gonococcus is found.

Treatment—Non-Operative.—The vulva and vagina, if infected, should be painted with silver nitrate solution, gr. xx to oz. j. With a sharp bistoury the cervix is bled. If the cervix is open enough the uterus should be washed out with either bichlorid solution, 1 : 10,000, or else with a saturated solution of boric acid. (See Septic Endometritis.) I prefer to use both solutions at the same sitting, the latter following the other. This treatment I repeat in twelve or twenty-four hours. In the interim I order douches of ¼ per cent. lysol at a temperature of 110°F. every two hours. If the case is seen early such treatment will often result in subduing acute symptoms after four days. But if the patient is not seen before the cocci have penetrated into the deeper portions of the endometrium, or if the cervix be markedly stenosed, the uterine washings are useless or can not be applied. The local blood-letting should be employed but once or twice, and the co-existing vulvitis and vaginitis will not require applications of silver solutions more than once a day. It is wise to pack the vagina with iodoform gauze wet in bichlorid, 1: 4,000. This dressing not only supports the uterus against painful jarring, but it also sterilizes the vagina as well as the discharge from the uterus. But it is not to be used in cases where intra-uterine washings are impossible, for there it will but dam up the discharges. Opiates will usually be found necessary for the relief of pain.

Operative Treatment.—Whether seen early or late after infection, I prefer to curette these cases. (See Curettage.) Certain teachers consider complications almost sure to result from such treatment, but the reason for these ill results will be found in the method of operating. I can see no reason for discriminating in favor of the gonococcus as against other pyogenic cocci. Certainly the worst that can result from operating is a complication, and this follows the let-alone treatment in nearly all instances. Indeed, it is doubtful if a general gonorrheal endometritis ever runs its course without producing some inflammatory process in either the tubes, ovaries, or pelvic peritoneum.

Sequelæ.—Gonorrhea of the uterus extends not through the lymph streams, but through the tubes—

extension by continuity of tissue. As a result we have salpingitis, ovaritis, and peritonitis the complications, and never see primary broad ligament abscess result. In view of the ravages which unchecked gonorrhea works in a woman's pelvis, the very first symptom must be vigorously met. The joint complications sometimes seen in men, I have never seen in women, but they do occasionally occur. Women who have had true gonorrheal endometritis and in whom it is not radically cured are always sterile. Those cases which do not produce some complication result in a chronic condition. My belief is that gonorrhea of the endometrium is *never* cured except by surgical means.

CHRONIC GONORRHEAL ENDOMETRITIS.

Symptoms.—Chronic gonorrheal endometritis is about the happiest result to be expected from an acute attack. I do not find the chronic state other than as a result of the acute. There is slight enlargement of the uterus in some cases; but old cases who have had repeated attacks, have usually small, hard uteri. The cervix in all is usually the seat of folliculitis with erosions. There is always a purulent discharge. The condition *may* exist without pelvic complications, *but I have never seen it*. Uterine pain is not present, but where the uterus is persistently enlarged there is a sense of weight in the pelvis. Whenever in such cases there is marked pelvic pain, it is an absolute indication that the causative acute attack has resulted in damage to the periuterine tissues. As a result of the aggressiveness of pelvic surgery a good many supposed conditions have been properly eliminated from our pathology, and many apparently innocent states of the uterus have been found to be accompanied by pronounced disease in the adnexa. As I said above, I do not find chronic corporal gonorrheal endometritis without complications. It is these complications so difficult to detect which produce the distressing symptoms and not the chronic inflammation in the uterus. These women are generally sterile.

Diagnosis.—For clinical purposes chronic inflammations of the uterus are characterized by one prominent symptom—a purulent discharge. There being no evidence of adnexal disease, we may put the case down as not due to gonorrhea. But far more difficult is it to determine whether the cervix alone or the entire endometrium is involved. To settle this the endocervicitis must first be cured. If this is easily accomplished, the flow of pus checked, and relapses without apparent cause do not occur, we can be sure that the pus did not come from above the os internum. But where the cervix remains inflamed under persistent treatment, or where the purulent discharge continues after the cervix is brought to a normal condition, we may know that the endometrium is involved. The persistence of the discharge despite energetic measures applied to the cervix convinces us that the corporal endometrium is inflamed. The patient's word cannot be relied upon in determining this, for she probably douches and washes away discharges. A piece of cotton large enough to fill the vaginal vault is wrung out in Thiersch solution and applied over the cervix. It is kept there by vaginal tampons. In twelve hours it is removed and the amount of pus discharged in that time can be determined.

Treatment.—The presence of adnexal disease is no bar to the methods of treatment. If the cervix be sufficiently open for the purpose, the uterus may be washed out (see Septic Endometritis). But as all the cases of chronic general gonorrheal endometritis which I have met have some degree of adnexal disease, I advocate curettage and the cul-de-sac operation (see Cul-de-Sac Operation). A curettage alone undoubtedly checks the source of infection, and, following it, some repair ensues in the inflamed adnexa. But we must consider both intra-uterine washings and curettage as merely palliative. If a radical cure is to be effected, the adnexa must be directly treated through the cul-de-sac. For a long time it has been my belief that chronic gonorrheal endometritis is never found except as a result of an acute process, and in this respect gonorrhea of the uterus differs from septic endometritis which may from the first be devoid of acute symptoms.

In no form of purulent endometritis do I ever make *applications*. These, while destroying pyogenic cocci, also kill the superficial cells of the endometrium and furnish no means for the escape of the destroyed tissue. The history-books of all of us bear many cases where acute pelvic inflammations have resulted from intrauterine applications.

TUBERCULAR ENDOMETRITIS.

Tubercular disease of the cervix is exceedingly rare, the disease being usually limited to the body of the organ. But it is occasionally met with, and then is secondary to vaginal tuberculosis. Occurring in the cervix the disease is either miliary or ulcerative, and is not often diagnosticated without the aid of the microscope; the miliary tubercles being mistaken for small cervical cysts and the ulcers for carcinoma.

It is rare to find the cervix and corpus uteri involved in the same individual; the uterus being affected secondarily, the cervix from the vagina, the endometrium from the peritoneal face of the uterus.

Symptoms.—Tubercular endometritis produces profuse leucorrhea which may contain caseous masses. The uterus is enlarged. Where menstruation occurs it is irregular or profuse, but the concomitant cachexia generally produces amenorrhea in the later stages. Otherwise the symptoms are those of chronic endometritis, *plus* the general symptoms of general tuberculosis where that exists.

Diagnosis.—Without finding the tubercle bacillus, a positive diagnosis is impossible. It is not necessary to excise portions of ulcerating tissue; the discharge will show the bacillus.

Treatment.—Whenever a diagnosis can be made, ex-section of the affected portion is indicated if it possibly can be done. The cervix without a tendency for the disease to extend upwards, may be amputated. But palliative treatment can not be applied to the corpus uteri, and curettage alone will not check the disease. Total vaginal extirpation of the uterus and adnexa is indicated, both because the uterine disease is commonly secondary to adnexal tubercular disease, and because extirpation of the tubercular uterus works such marvellous changes in the metabolism of the blood as to hold in abeyance for years tubercular lung disease. Because of local and general reasons, if I may so term them, hysterectomy is indicated in corporal tuberculosis. Any other operative procedure but plays with the disease, opening new channels for its extension. Local treatment is useless. Excision is the treatment for tubercular disease of the genitals. I have seen a phthisical woman gain thirty pounds in two months after this operation. No one who has not estimated the quantity, often ounces a day, of discharge coming from these women can have any idea of the drain upon their systems.

The failure of agents lauded as corrective of tuberculosis to even modify the disease when the organ is within easy access and the treatment applied under the eye, is a commentary upon the methods of reasoning of those who advocate them.

Sequelæ.—When occurring as a primary disease, uterine tuberculosis will surely extend to the peritoneum and adnexa, to be followed by general tuberculosis.

PELVIC INFLAMMATION.

Up to a few years ago it was undoubtedly the practice with most surgeons to remove through the abdomen all ovaries and tubes which presented evidences of inflammation, whether these were diseased primarily, or as the result of a lymphatic infection in the pelvis. So long as this remained the established rule of procedure in dealing with tubo-ovarian disease, precise differentiation of the various lesions was not necessary. But a more careful study of the manner in which the gross lesions were produced, together with the application of those general surgical principles which govern the treatment of inflammatory lesions elsewhere, has taught us the necessity for carefully separating those lesions which necessitate the removal of the diseased organs from those which are relieved by conservative measures. It therefore becomes our duty to enter into a thorough analysis of each case. To do this it is not essential to a proper conclusion that a bacteriological examination be made, but the correct treatment can be reached by studying the clinical history of each case. Although the ovaries and tubes, as well as the pelvic peritoneum, will probably suffer in most cases where the infection passes outside the uterus, yet all the structures will not be equally damaged. The manner in which the infection reaches the pelvic structures as well as its nature will indicate somewhat the organ we will find most affected. In all infections, for instance, occurring in the uterus pregnant after the third month, those lesions which result from pelvic peritonitis are to be expected, for in such a uterus the lymph streams and not the tubes are the chief carriers of the infecting agents. And inasmuch as the poison of gonorrhea travels not through the lymphatics, but through direct continuity of tissue along the uterus and Fallopian tubes, this may usually be eliminated as a cause of the trouble, and the case be set down as due to sepsis. As a result we have produced the effusion of lymph on the peritoneum, suppuration in the folds of the broad ligaments, or ovarian abscess, one or all. (Secondarily salpingitis may result.) Conversely, in a uterus with undeveloped lymphatics, that is in the unimpregnated uterus, the invasion is usually through the tubes, and may be either gonorrheal or septic. The resulting lesions are salpingitis with secondary involvement of the peritoneum and perhaps of the ovary. Furthermore the results of these various kinds of infection may be of a purulent, a cystic, or even a connective-tissue type. To characterize all these lesions as merely "tubo-ovarian disease" is to subject them all to one method of treatment, removal.

The diffuse peritonitis due to abortion is cured in the same way as the pleurisy due to rib necrosis, by scraping away the infected focus; cysts of retention, as cystic ovary and hydrosalpinx, are to be treated as similar accumulations elsewhere, that is by evacuation. Purulent accumulations here behave exactly as elsewhere in the body. If found in preformed sacs, as the tubes, they are cured by removal or by evacuation and obliteration. If seated in the continuity of tissue, as in the ovary or broad ligament, evacuation will cure. But the situation of the diseased focus, whether high up or low down, will influence the method of operating; for the element of drainage bears a most important part in all such work. I mention these facts briefly to emphasize the importance of carefully considering each case upon its merits, and to show the utter folly of laying down hard and fast rules for dealing with the several lesions. We formerly heard much about "cellulitis." Cellulitis does not occur in the tubes, that is, suppuration is not found in the tubal walls except as consecutive upon more marked suppuration in their cavities. Suppuration between the folds of the broad ligament is not cellulitis. The pus is not produced in the thin cellular layer of the ligament, but in its lymphatics. Broad ligament "cellulitis" is broad ligament lymphangitis. The only true cellulitis found in the pelvis is that which is seen in the ovarian stroma. There may be effused about the broad ligament a large mass of lymph as a result of lymphangitis between the folds of the ligament, and suppuration in such an accumulation may simulate "cellulitis." Even when applied to ovarian suppuration, the term is misleading. It should have no place in our nomenclature of pelvic lesions, and I do not accept it as descriptive of any lesions I have ever seen in the woman's pelvis. Migration of cells takes place in the pelvic structures because of invasions from without. If the cells die, pus is produced. The generalization "cellulitis" is meaningless. Pyosalpinx, ovarian abscess, pelvic peritonitis, are precise statements of gross lesions, and gross lesions, not microscopic changes, are the basis for our operations.

The great advance made in the treatment of pelvic inflammatory lesions is due to our knowledge that they are produced by exactly the same agents as cause inflammations elsewhere. Cold feet, fright, menstruation, etc., have ceased to be causes for pus in the pelvis, although used as excuses for unclean surgical operations and infidelity.

Certain germs produce certain lesions under certain conditions. The pathology of pelvic inflammations is as accurate as any other, and upon an accurate knowledge of the causative germs, the conditions under which they exist, and their paths of extension, will depend the proper treatment of the results of their activity.

This precise differentiation of the conditions under discussion cannot always be made upon examination, but the cul-de-sac exploration will clearly show them.

Starting, then, with the history of the case and the data furnished by carefully kept clinical notes and repeated examinations, we are in a position to determine the propriety of operative interference. And the surgeon will decide whether his attention should be limited to the uterus alone or whether the adnexa should be attacked. In the latter event the exploratory part of the operation can be made as satisfactorily through the vagina as through the belly and with far less risk to the patient. Meeting with certain conditions, the vaginal section can be made a curative operation without removal of any organs and without risk. This much cannot be said of laparotomy.

All these questions are brought under discussion in the proper places.

Diagnosis.—Before attempting to proceed with any method of treatment, it is essential to determine that the case is one of inflammation of the generative organs. The diseases with which pelvic inflammation is most commonly confounded are: appendicitis, ureteral inflammation, acute cystitis, general peritonitis and suppurating ovarian cyst.

Appendicitis.—This cannot be mistaken for pelvic inflammation of course, except where the disease is on the right side. The pain of appendicitis is usually situated somewhere about a line drawn from the umbilicus to the anterior superior spine of the ilium. Pain due to tubo ovarian disease is commonly much below this. The sensitiveness to touch in appendicitis is greatest when pressure is made upon the belly over the cæcum, while greatest pain in adnexal disease is developed by vaginal examination. The pain due to appendicitis radiates upward toward the hypogastric region, while that due to tubo-ovarian disease extends downward toward the pubic region of the pelvic cavity. In appendicitis disturbed bowel function frequently precedes the attack, while it follows salpingitis, if it occurs at all. Symptoms of colitis often accompany appendicitis, while they are absent with tubal disease. Tympanites is usual in appendicitis, the distention being general, while in pelvic inflammation it is usually limited to the colon. In appendicitis there is absence of signs of genito-urinary inflammation, such as endometritis, urethritis, etc.; while some one of these is present with salpingitis. Very often tubo-ovarian disease coexists with appendicitis, and the inflamed organs may be matted together in one indistinguishable mass. In such cases the operator must consider the lesions as abdominal rather than pelvic. Appendicitis does not cause fixity of the uterus, and digital search in the vagina does not increase the pain of appendicitis. The contrary is true where pelvic inflammation exists. But appendicitis may coexist with pelvic inflammation. As the abdominal symptoms will predominate over the pelvic a suspicion of the appendicitis must be held. Such cases are to be viewed from the abdominal side if the cul-de-sac exploration fails to clear up the diagnosis.

Cystitis.—Ardor urinæ, blood or pus in the urine, pain on urination, suprapubic sensitiveness, infrapubic tenderness, mark cystitis. In cystitis vaginal examination does not develop pain except where the space *anterior* to the uterus is pressed upon. In pelvic peritonitis the *lateral* fornices are most sensitive. Again, in pelvic peritonitis there is not marked disturbance of bladder function, while in cystitis there is, and there is usually pronounced inflammation of the uterus in cases of peritonitis which which is not present in cystitis. The uterus, tubes, ovaries and peritoneum are uninvolved in cases of cystitis.

Ureteritis.—Inflammation of the ureters is usually secondary to cystitis. A careful urinalysis will show pus, ureteral epithelium and often blood in the urine drawn by catheter. There is the same pain in the lateral pelvic walls as is found in adnexal disease, but there is no fixity of the uterus, and moving the uterus with the finger does not increase the pain. Ureteral calculi cause spasmodic pain similar to that of acute salpingitis, but it radiates down to the urethra, produces bloody urine, and the spasm is often accompanied by involuntary discharge of urine. The pain of pelvic peritonitis is constant, and the fever produced by tubal inflammation is usually higher than that seen with cystitis and ureteritis.

Suppurating Ovarian Cyst.—There is the same mass, tenderness and fixity as accompany tubal disease. In fact, the symptoms produced by such a cyst are more due to the concomitant peritonitis than to the cyst. It is exceedingly difficult to differentiate between a large pyosalpinx or ovarian abscess and a small inflamed ovarian cyst. If the cyst is of sufficient size it will press the uterus upward and forward, whereas a large adnexal abscess will displace it laterally.

Haste in making examinations will lead the surgeon into many errors. I have seen a good man cut open the belly for typhoid fever in a woman with gonorrhea, thinking he had tubal disease to deal with.

From what has been said, it will be seen that the seat of pain and its character are invaluable in making a diagnosis. Therefore, opiates are not to be used until the diagnosis is clearly made, if at all. I have laid some stress upon the evidence of disease as developed by the cul-de-sac operation. This exploratory procedure being free from danger, and occupying but a few minutes time, can be resorted to where a positive diagnosis is necessary. Lesions being by it revealed, the removal of which would endanger life or is forbidden, the operator can not only retreat without having damaged his patient, but in most cases of pelvic disease, he will afford some measure of relief by that operation alone. Such can not be said of the abdominal section.

General Suppurative Peritonitis.—In some cases of diffuse pelvic suppuration, with sudden and sharp extension of the infection, it will be hard to differentiate between a local and a general infection of the peritoneum. General purulent peritonitis produces great shock. The pulse is generally above 120, the temperature ranging about 103°. Tympanites is pronounced, vomiting incessant; restlessness, sordes on teeth and lips, dry, cracked tongue, mental anguish, pinched face, muscular twitchings and delirium, mark general peritonitis. As each hour passes the symptoms become rapidly more grave. There is no remission. Without an examination, the experienced physician can, by carefully studying the symptoms, determine whether the case is one of localized lesions, with severe general symptoms, or one of general peritonitis. Examination will be of little service in making a differential diagnosis between local and general peritonitis. I have seen at ten in the morning a case, apparently, of bilateral pyosalpinx due to abortion, which when operated upon at three in the afternoon was almost moribund. Certain cases of virulent pelvic inflammation will, in a few hours, become general throughout the peritoneal cavity. This is especially true of cases of post-partum and post-abortum infection.

General peritonitis is one of the results to be expected from a policy of delay. The vomiting, rapid pulse, dry tongue, stupor and tympanites, which result from the use of opiates in a grave case of pelvic inflammation will often render a diagnosis of the extent of the infection utterly impossible. The cases where surgeons have opened the belly and found all the disease in the pelvis are many. The surest way to mask the symptoms of pelvic inflammation is to give opium.

Treatment.—OPIATES.—These I rarely use and for a number of reasons. To relieve pain by opiates is to mask the symptoms of extension and complications. The bowels become blocked, and this permits the migration of the intestinal germs; the stomach is deranged, and the bladder function disturbed. Brought to a case of pelvic inflammation, if he contemplates possible operation, the surgeon should withhold opiates. Other means are at his command to ease pain, and the relief afforded by opiates is but a borrowed ease, to be paid back by vomiting, tympanites, and spread of the infection.

INTESTINAL CLEANLINESS.—From the first I insist upon this. Placing the patient upon the left side with the hips elevated, I insert a Martin's tube (Fig. 13) and throw into the bowel a quart of tepid normal salt solution. Part of this is retained. To obtain retention the injection should be given at night, or when the bowels do not habitually operate. This cleanses out the colon, flushes the kidneys and allays thirst. I do this each night. The maneuvre is very simple. The outflow tube is corked up, and the fountain syringe is raised four feet above the bed. The water should be of a temperature of 103° to 105° F. These irrigations, by removing hard fecal masses, and relaxing muscular spasm, relieve pain and prevent tympanites. Purgatives I never use, but mild cathartics, as an ounce of Rubinat water every second morning in a glass of plain water an hour before breakfast, I employ.

There are but two ways of treating these cavity inflammations: the locking-up plan (opiates), and the eliminative. I prefer the latter. No man who has ever contrasted the two will accept the first.

DOUCHES.—Some patients express themselves as relieved of pain by douches, others dislike to be disturbed.

FIG. 13.—Martin's tube for rectal irrigation.

But they are demanded where there is much discharge. I prefer to use lysol ¼ per cent., at a temperature of 110° F., every four hours.

ABDOMINAL DRESSINGS.—Poultices cause effusions of blood beneath the skin and vary in temperature from 90° to that at which they are put on. I prefer to use an ice-bag over the pubes. It is easier handled and relieves pain, and tends to limit the peritonitis *a very little*.

FLUIDS.—I encourage these women to drink large quantities of water. A good plan is to administer three ounces each hour when awake. Ten drops of lemon juice added will keep the tongue clear. Milk, koumyss, etc., having poor food value, I do not use. Besides, it is difficult to get rid of the cheese left after such preparations. Alcoholics are never indicated. In women deep in infection, and with bad kidneys and thready pulse, I do intravenous infusion. Anesthesia is unnecessary for this. (See Article on TRANSFUSION.) Far better than all drugs is this procedure. It flushes the kidneys, eliminates toxins and stimulates the heart. The amount of urea excreted is increased and the albumen diminished. The effect is immediate.

DIET.—Half diet is indicated in mild cases. Farina, hominy, an egg, and a little coffee for breakfast; a little meat, as beef, chicken, chops, etc., with potatoes and cream at midday; chicken soup and toast at four o'clock; and squeezed beef-juice and toast at eight. But all the time an abundance of water.

To those very ill I do not give solids. In the morning, toast, and a very little coffee. Once every three hours after that, either two ounces of chicken broth or one ounce of squeezed beef-juice is given, each time with a little toast. *No milk.* To those who vomit continuously, I give nutrient enemata. Prepared foods, pre-digested foods and fermented foods are not to be used.

LOCAL APPLICATIONS.—Pelvic pain, due to inflammation, is diminished by ice-bags over the suprapubic region. Spasmodic pain is eased by poultices. To the vagina, 10 per cent. ichthyol in glycerin, applied by means of a syringe, will reduce pain and act beneficially upon the inflammation at all stages. Local blood-letting, applied to the cervix, is of great value in relieving pain and vascular stasis.

To me, the free purgation used by some is as bad as the opium treatment. All that is needed is a through and through stool once every day or so, but the colon must be kept empty. It is unwise to irritate by strong saline cathartics the thirty feet of intestinal mucosa. Sepsis in the pelvis is not checked by it, and shock is increased. Perfect rest in bed is imperatively necessary. The non-operative treatment of pelvic inflammations seeks: the improvement of the tissue-resistance; limitation of the infecting agent; and maintenance of the general strength while the invaded organs are overcoming the infection.

SALPINGITIS.

Gonorrheal Salpingitis	}	Acute; Chronic.
Septic Salpingitis	}	Acute; Chronic.
Tubercular Salpingitis	}	Chronic.

ACUTE GONORRHEAL SALPINGITIS.

The causative germs are gonococci. These may have entered fresh from an acute clap in the male, or from a gleet, or may be the result of a rekindling of the spark which has for some time been dormant and latent in some other part of the genito-urinary tract of the woman (Fig. 14). The disease is usually bilateral, though differing often in severity upon the two sides. The first stage is one of congestion and edema. When the infection reaches the fimbriated end, a local peritonitis results. The peritonitis is secondary to the salpingitis, because the gonorrhea travels through the tubes and not through the lymphatics. The fimbriæ of the affected tube turn in and their peritoneal surfaces cohere, owing to a circumscribed peritonitis about the fimbriæ. Thus the tubal contents become locked in, and, secretion continuing, a cyst of retention is formed (Fig. 15). Usually this latter is a *pyosalpinx*, occasionally a *hydrosalpinx*. Or, the secreted fluids may become almost wholly absorbed and the condition be marked by the production of new connective tissue in the walls of the tube, which, contracting, constitutes *tubal sclerosis*. Sclerosis with hydrosalpinx, or with pyosalpinx, is very common.

In the first or acute stage, the tube is deeply discolored and easily torn. It may measure as much as an inch in diameter. Upon section the lumen is found not much distended and the increase in size is due to submucous infiltration.

The tissues are so swollen that the rugæ are almost

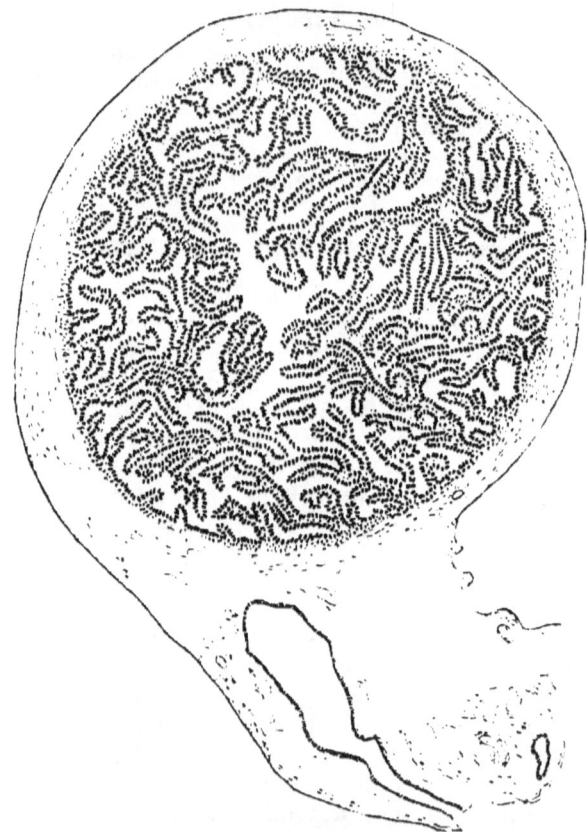

FIG. 14.—Section of the normal Fallopian tube near the abdominal ostium (Beyea).

obliterated as separate folds. The uterine end of the affected tube is still patent, and the creamy purulent contents of the tube escape into the uterine cavity.

SALPINGITIS.

After a time the uterine end of the tube is also closed, and there is formed a permanent fluid accumulation in the tube. The tube may be knotted and twisted or sausage-like. Spontaneous cure is rare, the case usually resulting in some permanent deformity. The heavy tube is prone to sink down behind the broad ligament. Commonly the adjacent ovary as well as the pelvic peritoneum is involved.

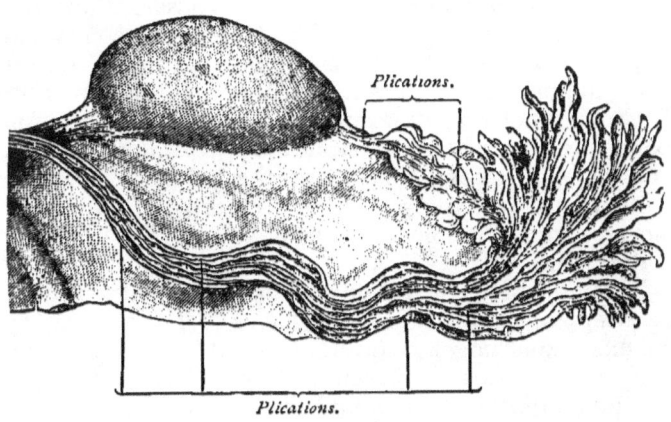

FIG. 15.—Longitudinal section of Fallopian tube, exposing the complicated longitudinal plications of the mucosa which expand into the fimbriæ (Sappey).

Symptoms.—Added to the symptoms of acute gonorrheal endometritis we have certain others. In addition to the uterine pain and cramps are pains originating in the neighborhood of the tubes and shooting down the thighs or upward. There is also pain in the sacroiliac region. Spasm of the affected part is common, I find, before peristalsis is stopped by the intensity of the inflammation The temperature (rectal) may reach 103° F. and the pulse nearly 120, but commonly the pulse is below 110. There is often vomiting and nausea. Tympanites and costiveness are common. The symptoms are most severe before the stage of stasis, that is, up to the fifth day. They then usually begin to subside somewhat and the case resolves into some one of the chronic forms. Upon vaginal examination the lateral pelvic walls are found tense and the uterus fixed by plastic effusion and tubal enlargement. The lateral fornices are exquisitely sensitive and a thorough examination may be impossible without narcosis. In seeking the enlarged tube, the viscera above are supported by a firm *steady* pressure of the palmar surfaces of the fingers applied over the ovarian region. One finger is inserted into the vagina and the cervix found. The finger is then crowded high up alongside the uterus, and as it is swept out toward the pelvic wall it will feel the tube roll over it. The tube is not very movable, and where marked fixation exists the finger can very accurately determine its size. The physician should stand at the patient's side, she being in the lithotomy posture. Placed in this way he can pronate and supinate his hand so as to feel all parts of the pelvis.

Upon opening the cul-de-sac there is usually at once an escape of serum containing flakes of lymph. When the examining finger is inserted (see cul-de-sac exploration) it appreciates the presence of recently effused lymph, for the rectum is often found attached to the posterior surface of the uterus by tender union, and the inflamed tubes are sealed to whatever structures they rest against, usually the broad ligaments. If the affected tube is low down, it is felt to be a rounded firm yet elastic mass, extending from one cornu of the uterus to the ovary or broad ligament. To one of these it is usually attached at its fimbriated end. At the cornu the tube is small, but as the free end is reached it is found to enlarge into a blunt knob. The false union between the tube and other structures is easily severed by the finger, and this produces slight parenchymatous oozing. Upon freeing the tube and ovary from abnormal attachments they can be brought into view at the vaginal vault. To accomplish this it is better to introduce a gauze pad above the affected tube and then draw down the tube below the pad. In manipulating the tube the finger only should be used, as instruments tend to break the swollen and friable structure. The gauze pad keeps the intestines above the vagina, but if these continue to prolapse into the vagina the table may be lowered. Upon inspecting the tube it will present the signs of intense inflammation described at the first of this chapter.

Diagnosis.—There are nearly always present other evidences of gonorrhea, and in most cases the pus from the uterus shows the gonococcus to be present. The various *pelvic neuralgias* are devoid of the symptoms of acute inflammation, general as well as local.

Ectopic gestation in the early months may present the same symptoms of pain, a mass, and tenderness; but in ectopic gestation there are not acute purulent endometritis, erosions, fever, etc. Inspection through the cul-de-sac will positively determine the question of the existence of acute salpingitis.

Treatment.—As soon as a gonorrheal process has passed outside the uterus involving the tubes and pelvic peritoneum, the disease has progressed too far to be cured by any form of treatment applied solely to the uterus. It is necessary to attack the complications. The uterus is curetted (see curettage). After this is done the posterior cul-de-sac is opened. If this exploratory incision shows that both adnexa are involved, one of two courses may be taken: either break up all the adhesions, split open the tubes (see conservative treatment) and pack the pelvis with gauze, or also make a vaginal ablation. Merely opening the tubes and swabbing them out with iodoform gauze will suffice in cases in the acute stage and before pus has become encapsulated within them. Of course the pelvic peritoneum is exposed to the dripping from the tubes, but this is caught by the gauze. If distinct pus sacs have already formed, they are to be treated as described under chronic inflammation. My experience teaches me that all cases of gonorrheal tubal disease at some time either relapse or become reinfected. To always do an ablation is to enter the dangerous field of preventive surgery. I prefer in most cases of first attack to perform the palliative operation, aiding repair by evacuation and drainage in the acute cases, and in the more chronic seeking the obliteration of the tubes by connective tissue hyperplasia; and then if a relapse occurs I ablate. Necessarily the woman's surroundings, position in life and her general condition will greatly govern my action. The palliative operation in the worst cases enables the operator to remove the case to another class: those in whom an *elective* radical operation can be performed, with all that that means to her and the surgeon.

If upon making the exploration through the posterior cul-de-sac, I find one tube alone involved, and the indication is apparently clear to remove it, I would preferably perform the palliative operation (which see); or else, not do a laparotomy, but ablate the uterus and both adnexa through the vagina. The relapses are so frequent where one tube is removed through the abdomen, the remaining one becoming infected often before the woman leaves her bed after the first operation, that I cannot lend my endorsement to the practice of abdominal unilateral salpingo-oöphorectomy in these cases. I am clear in my conviction that we must either be wholly conservative or most thorough in our treatment. (See conservative treatment).

If we could only keep these women away from men, the ideal operation, one with permanent results so far as prevention of further suppurative processes is concerned, would be the palliative operation through the cul-de-sac. But the unfortunate creatures will return to the husbands or lovers who infected them.

Sequelae.—The neglected cases go on to the formation of either a sacculated pyosalpinx through closure of both ends of the tube, or a sclerosis, or to a chronic relapsing type; rarely does hydrosalpinx result from gonorrhea.

If the currettage and cul-de-sac operation is done in a first attack of gonorrheal salpingitis, return of the tubes to a normal state is possible. I have had one patient on whom this operation was done report herself pregnant.

But where gonorrhea has been allowed to proceed unchecked to the formation of a pyosalpinx, and where repeated attacks have occurred, which, while not producing pyosalpinx, have wrought permanent lesions in the walls of the tubes, the cul de-sac operation is merely palliative, and the restoration of the tubes is but partial.

ACUTE SEPTIC SALPINGITIS.

The usual causative germs are staphylococci or streptococci (Fig. 16). Sometimes these reach the tube by extension along the lining of the uterine and tubal lining membranes. In most cases, however, the tube is involved subsequently to the pelvic peritoneum, the infecting process first reaching the pelvis through the lymphatics which extend from the uterus between the broad ligaments to the iliac glands. For this reason we most often find this form of infection following interrupted or completed gestation, for in this state the lymph streams are sufficiently developed to carry the infecting agent. The infection is more apt to affect one tube than is the case with gonorrhea. The lesions induced in a tube so inflamed are identical with those which follow gonorrhea.

The appearances, both gross and microscopic, are the same, except regarding the causative cocci.

Symptoms.—Very often there is a chill ushering in the septicemia. The temperature rises rapidly and the pulse is disproportionately high. For example, the temperature may be but 102° and yet the pulse range from 120 to 140 beats a minute. Local pain is severe, at first spasmodic in the uterus, and then becoming continuous, as the general pelvic cavity becomes involved. Altogether the clinical picture is not significant of the cause of the infection *except* that it is more commonly found following gestation or unclean operations, and that the examination may reveal the peritonitis occurring before tubal enlargement can be appreciated.

It is never possible without the use of the microscope to differentiate positively the cause of the inflammation.

But given a history of an operation upon the uterus, or abortion, or labor, followed by fever, pain, peritonitis,

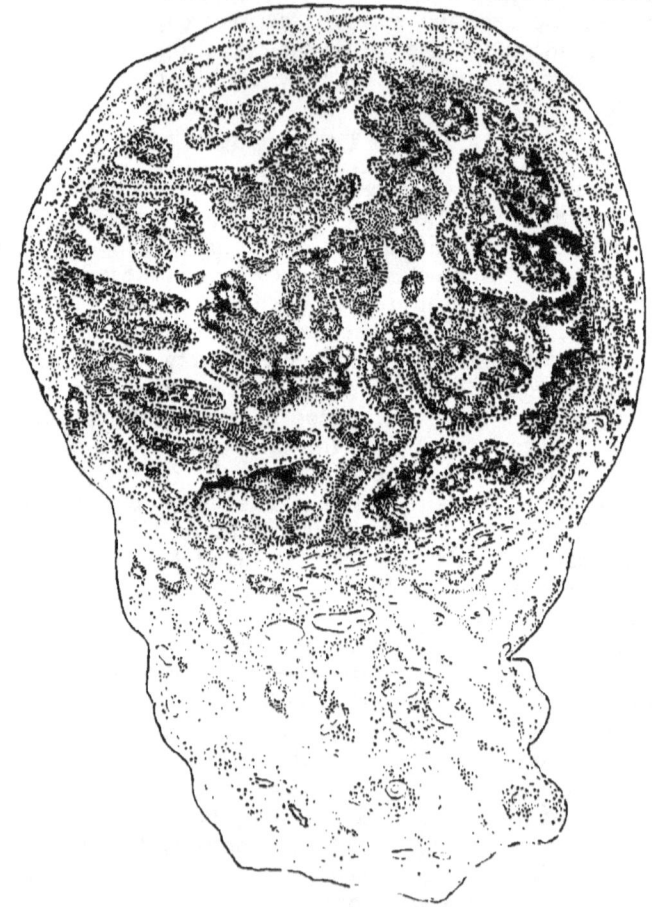

FIG. 16.—Acute septic salpingitis. Note the cellular infiltration in the walls of the tube as well as in the plications of the mucous membrane. Section about the middle of the tube (Beyea).

and later a swollen tube, we may infer that the case is septic and not gonorrheal. This inference becomes a strong probability when we fail to find signs of gonorrhea elsewhere, notably in the urethra and vulvo-vaginal glands; and the diagnosis is rendered certain by the microscopic examination of the discharge. But the possibility of mixed infection must be remembered, for all pyogenic germs may be in the same case. Upon bimanual examination the same signs are found as exist with salpingitis due to gonorrhea.

The cul-de-sac incision and examination do not aid in differentiating the cause of the salpingitis, and the affected tube presents the same adhesions, discoloration, swelling, etc., as are found in gonorrheal infection. But there is apt to be more peritonitis with sepsis than with gonorrhea.

Treatment.—UNIMPREGNATED CASES.—The moment the tube or the peritoneum shows evidences of being involved, the uterus should be curetted. If this is done before there is actual death of cells with the formation of pus, a cure will be started in the vast proportion of cases, no matter how extensive the peritonitis may be. The first indication for treatment of a septic uterus with complications short of the formation of pus is the removal of the causative focus, the infected endometrium or decidua. If a plastic operation has caused the trouble, all sutures must be ripped out and the uterus cleansed by irrigation with boric acid. The raw surfaces of the wound should then be painted with pure carbolic acid.

By employing this curettage early more women have been saved needless mutilating operations than by any other treatment. Suffice it to say that the ability of the tube to recover after the causative focus is removed by curettage is far greater when sepsis starts the inflammation, and less when gonorrhea is the cause.

Still here, as in the article on gonorrheal salpingitis, the rule holds good to open the cul-de-sac in addition to curetting in all cases of relapsing septic salpingitis and in those seen after labor or abortion, and always when evidences of pus formation are present. The combined procedure removes the diseased endometrium and drains away the complications in the pelvis. Enormous quantities of muddy toxins escape through the cul-de-sac gauze in such cases, showing to what extent the lymphatics are drained. The technic of this incision and the method of dressing these cases are found in an appropriate chapter.

If I find one tube only involved to a degree which would seem to warrant its removal, I always apply conservative treatment to its cavity, and do not debate the propriety of its removal. If I find both tubes acutely inflamed in a first attack I do not remove them or consider ablation; for I have seen apparently hopelessly involved tubes completely recover when freed and drained. But in cases many times before infected, and in those relapsing after conservatism has failed to cure, the repair power is so damaged that it is unwise to attempt conservatism. I therefore advise the radical vaginal operation. From the difference between the treatment applied to the gonorrheal and the septic cases will be seen the importance of a differential diagnosis.

CHRONIC SEPTIC AND GONORRHEAL SALPINGITIS.

Unchecked septic and gonorrheal infections always leave the invaded tube permanently damaged. If the tissues have marked resistant power there may be merely a production of new connective tissue. A repetition of the infection results in more scar tissue, until in time the tube and ovary become but masses of cord-like connective tissue, constituting the condition known as *pachysalpingitis* or *sclerosis* (Figs. 17, 18). This is commonly bilateral, and the process frequently involves the uterus. Or, for some unknown reason, the infection merely causes occlusion of both the fimbriated and uterine ends of the tube. The tubal secretion is retained, and the tube becoming progressively enlarged, hangs at the cornu like a large grape, with thin walls and clear contents. Sometimes a combination of sclerosis and retention is met with, forming a sacculated hydrosalpinx (Fig. 19). I have found much

SALPINGITIS.

the larger number of hydrosalpinx cases to be due to mild sepsis after abortion and labor. The explanation is that sepsis here closes the fimbriated end of the tube, and inflicts but slight damage upon the tubal lumen.

The graver results of pelvic inflammation are pyosalpinx and ovarian abscess. Very frequently we find them

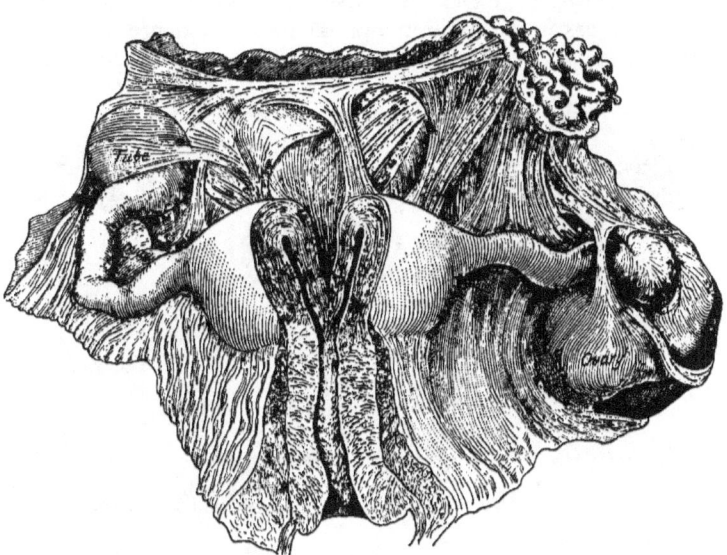

FIG. 17.—Chronic salpingitis with general adhesions of tubes, ovaries, and uterus (Bandl).

associated in the same case. The tube is filled with pus, its walls thickened, and recent lymph, as well as old adhesions, are found upon its surface. The adhesions are very dense. If the ovary be purulent, there may be one main sac with the ovarian stroma riddled by communicating sinuses; or the pus may be in isolated localities. These organs are always firmly attached to the adjacent structures. The longitudinal plicæ of the endosalpinx are obliterated, and the cavity of the tube is lined by a roughly granular membrane, " pyogenic membrane."

PELVIC INFLAMMATION.

Sclerosis.—There is usually a history of many attacks of endometritis, and many cases have been subjected to repeated curettages. The menses are diminished. Intermenstrual leucorrhea is present, but is not profuse. There is continuous pelvic pain, which at irregular intervals is exacerbated. There is no fever, and no evidences exist of more than the local distress. The pain is severe; in many cases the women appear to suffer more than do pus cases. The effort to menstruate from the sclerosed

FIG. 18.—Removed by vaginal ablation. Pachysalpingitis or sclerosis.*

endometrium produces great pelvic tenesmus. Most of the cases are stout. Fever and acceleration in pulse rate are absent. The condition is exceedingly common in old prostitutes. Upon examination the uterus is found atrophied and high in the pelvis; it is fixed there by adhesions and not by plastic or other masses in the lateral fornices. The sides of the vagina are drawn to the cervix by contraction in the pericervical tissues. The tubes can be felt as cords extending from the cornua up to the pelvic brim. The ovaries cannot often be felt. Commonly the uterus is diminished in size, but often cases are met with where the organ is enlarged (Fig. 20). There

* The wire screens used in these illustrations are all of half-inch mesh.

FIG. 19.—*a, b, c,* Uncollapsed walls of a left hydrosalpinx; *d, e, f,* a right sacculated hydrosalpinx; *g,* the obliterated end of the right tube. A case for conservative operation (Winckel).

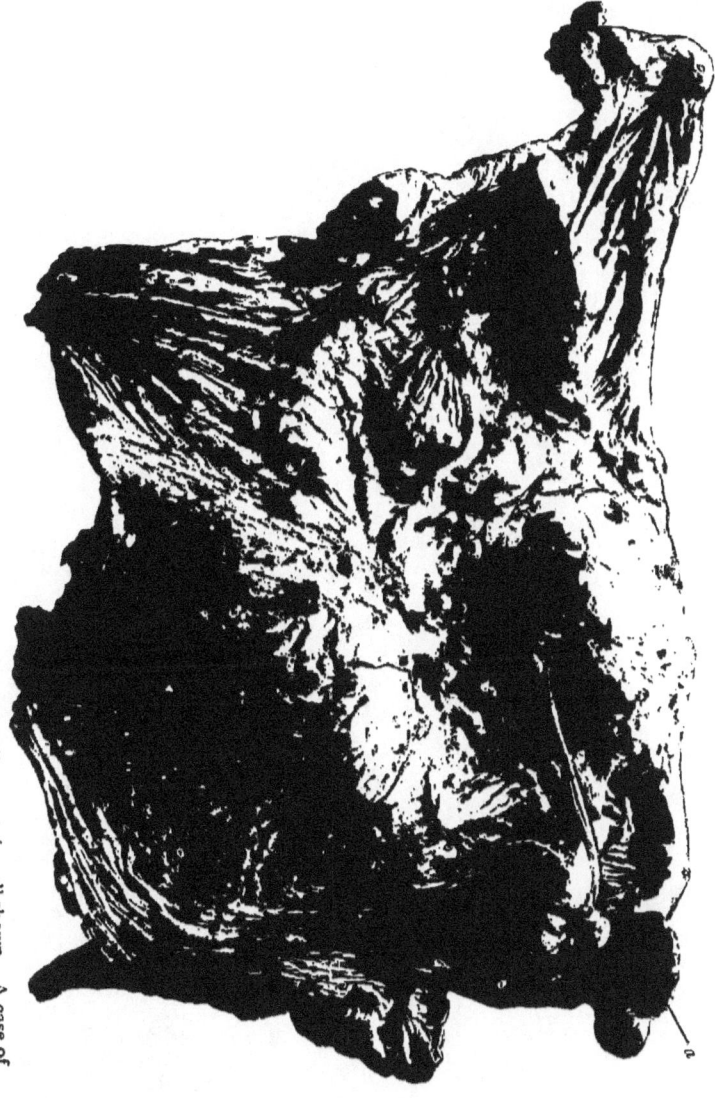

FIG. 20.—*a*. Fimbriated end of tube. The distortion of the tubes and broad ligaments is well shown. A case of sclerosis. Not to be relieved by conservative treatment (Winckel).

SALPINGITIS.

is a fixity about these uteri, without evidences of effusion, which is characteristic. The pain is produced, not by inflammatory tension, but by constriction of the nerves by connective tissue. Nature has employed her surest method of obliterating the affected organs, that is by connective tissue hyperplasia. These women are always sterile (Fig. 21).

Opening the cul-de-sac is difficult. The tissues are firm and the scissors must be used freely. Even when entered, not much space can be secured, owing to the contracted vaginal vault. There are commonly felt firm old bands between the contents of the pelvis. The tubes

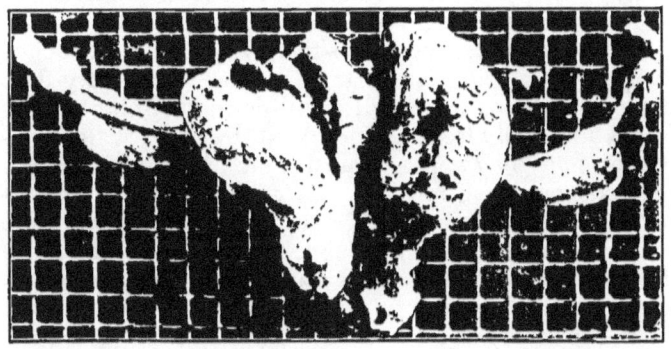

FIG. 21.—A case of genital sclerosis with hypertrophy of the corpus uteri. From an old prostitute who had been repeatedly aborted. Vaginal ablation.

can be traced to the lateral pelvic walls, or lower, behind the broad ligaments, as hard cords. They are freed with difficulty; in fact, sometimes it is utterly impossible to release them at this point. Upon direct inspection they are not pink as in health, nor brawny as are pus tubes, nor livid as in acute salpingitis. They appear as firm, pale cords, sometimes nodular. The ovaries are shrunken and scar-like.

Treatment.—A cure can be effected by ablation only. Ichthyol 10 per cent., on tampons or injected into the vagina, sometimes relieves the pain. The lesions are permanent and progressive. The cul-de-sac incision and conservative treatment afford no relief.

Hydrosalpinx.—As most of these cases follow abortion or labor, there may be elicited a history of perhaps mild infection at that time. There are not repeated attacks of peritonitis arising from the tube, but, of course, an affected endometrium may give rise to them. Still, as a rule, the course of a case of hydrosalpinx is more free from attacks of peritonitis than are pus cases. There is no fever and no continuous pain. Over-distention of adherent bowels produces pain in the tubal locality. Women may have large dropsical tubes and suffer but little. They are very commonly felt when examination is made, because of other conditions, as retroversion. The masses are not very sensitive, are not firmly attached by lymph, and communicate the sensation of very fluid contents. Upon opening the cul-de-sac they are readily found and easily freed (Fig. 22). Presenting at the vaginal incision they appear translucent and opalescent, or perfectly clear. Their sacs are transparent and exceedingly thin. They are easily ruptured by handling, and can be confounded with subperitoneal cysts only. Of course they are attached at the cornua, and may exist as single large cysts or as sacculated bunches of separate cysts occupying the tube cavity (Fig. 23).

Treatment.—They are to be treated by opening the cul-de-sac, freeing the sacs, and incising them with scissors. As the clear sterile fluid escapes, it is caught by gauze. The affected tube should be slit open for an inch. It is not necessary to do more. The tube oozes but little after incision and is returned into the pelvis. Preferably the incision should extend from the fimbriated end along the top of the tube. After the operation is finished the cul-de-sac is plugged with gauze which extends just within its cut edges. The first dressing is made in about eight days. No fever follows the operation. These cysts do not call for removal. (See conservative cul-de-sac operation.)

Pyosalpinx.—This is a purulent cyst of retention (Fig.

FIG. 23.—Showing a hydrosalpinx projecting into the vagina through a cul-de-sac incision.

SALPINGITIS.

24). The tube is more dilated near its fimbriated end, and at the cornu of the uterus it is quite small and hard, and its lumen obliterated (Fig. 25). Very often the pus tube is associated with an ovary of normal appearance, but in most cases the ovary also is involved in the mass of inflammatory products, sometimes producing a tubo-ovarian

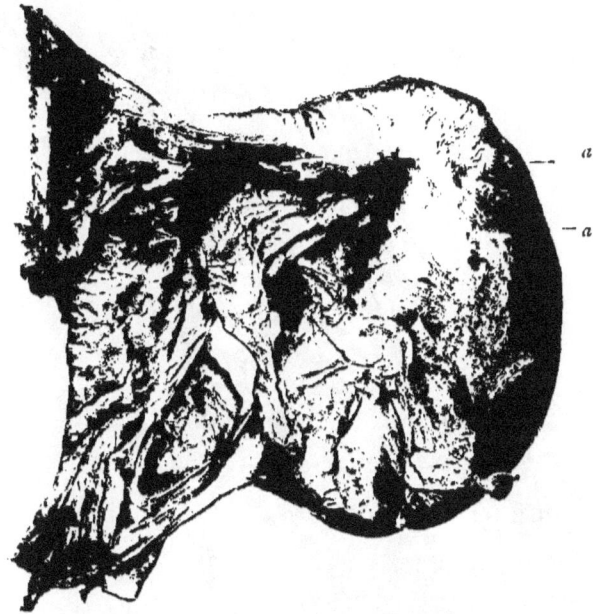

FIG. 22.—*a, a,* Hydrosalpinx. A lesion readily relieved by conservative operations through the vagina (Winckel).

abscess. There are commonly many adhesions between the tube and adjacent organs. Some of the adhesions exist as old bands, but recent lymph is generally always found.

Symptoms.—These are essentially those of acute gonorrheal or septic salpingitis in some cases, and in others there are no subjective symptoms other than a sense of moderate discomfort. Beyond a history of an infection, with possibly the presence of some evidence of gonorrhea, there are no symptoms different from those found

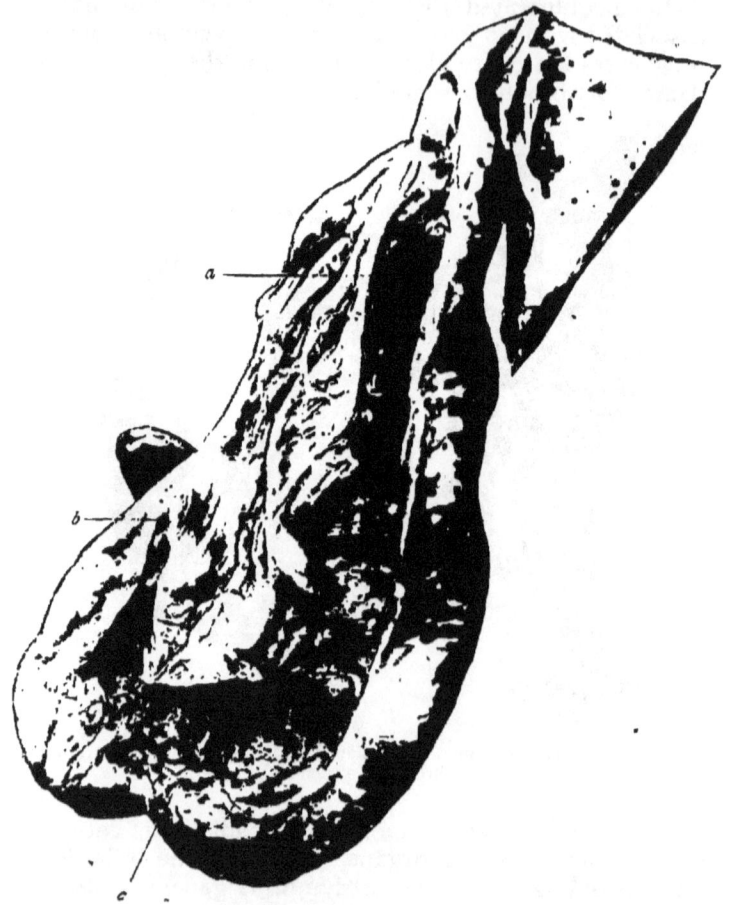

FIG. 24.—An old pyosalpinx. *a*, The thickened tubal wall; *b*, the occluded fimbriated end; *c*, the tube split open, showing the "pyogenic membrane" (Winckel).

in other suppurative processes in the pelvis. Purulent endometritis commonly coexists.

SALPINGITIS.

Upon examination of a case of salpingitis which has gone on to the formation of a pyosalpinx, the uterus is found more or less fixed. On the infected side is felt a hard yet fluctuating mass. This may be low down in the cul-de-sac, or high near the pelvic brim; usually it occupies a position somewhat below the level of the normal tube. It is sensitive upon pressure, the pain produced depending largely upon the acuteness of the stage of the inflammation. The mass is felt to be pedunculated, that

FIG. 25.—Bilateral pyosalpinx. Vaginal ablation.

is, there is a distinct sulcus between it and the uterus. This is not always so, and the distended tube may be closely matted to the posterior and lateral walls of the uterus by plates of dense exudate (Fig. 26). In size these pus-cysts may reach the dimensions of the pelvic cavity, crowding the uterus deep into the vagina, or high up upon the brim to one side. Fluctuation can nearly always be detected. In eliciting this it will be necessary to firmly support the tube by pressure from above while the vaginal finger determines the fluidity of its contents. Usually the disease is bilateral, the common association being a larger tubal abscess upon one side and a purulent salpingitis on the other. But both tubes may be of equal size. If the sac communicates with the gut, profound

86 PELVIC INFLAMMATION.

septicemia may set in, with high temperature and quick pulse. Even in those cases devoid of acute symptoms the evening temperature is higher than the morning.

Upon opening the cul-de sac, the finger appreciates the presence of dense adhesions. At once there escapes a variable quantity of serum. In old cases the cul-de-sac may be entirely obliterated and the finger be unable to enter except high up on the posterior surface of the uterus. Sometimes the first thing felt will be a knuckle

FIG. 26.—Right pyosalpinx. Left pyosalpinx. Vaginal ablation.

of small gut or of omentum. If the pus tube be low down, the finger will reach up unobstructed for a short distance, and then, being swept out to one side, will feel the mass. It is adherent behind the broad ligament, elastic but firm, as though there were fluid locked within a thick capsule. With the finger circling the periphery of the tube, it is gently freed below and will then be felt attached to the cornu of the uterus by a narrow neck. These tubes vary greatly in size and position. Direct inspection reveals a discolored sac, with bleeding points where adhesions have been severed, and flakes of lymph covering the surface. There are points of deep injection. If there be an ovarian abscess alone or coexisting, it can be differentiated from the tube only by tracing its attachments. Usually an ovarian abscess is attached higher than a pus tube, through the influence of the infundibulo pelvic ligament. The appearance of each is much the same. Ovarian abscess is usually firmer and more rounded than a pus tube. Very often the finger immediately upon penetrating the peritoneal cavity will meet with two deeply prolapsed tubes, the peripheries of which are agglutinated. The peritoneum of the cul-de-sac is much thickened and often must be severed by scissors. The evidences of intense inflammation, its attachments and shape, usually that of an elongated pear, will determine the character of a pus tube. If the trocar be plunged directly into the mass the diagnosis will be established by the escape of pus, often greenish and stinking, or even blood stained.

In tubercular pyosalpinx, tubercles may often be seen upon the surface of the sac. Upon the right side very often the tube is attached to the vermiform appendix, and at all points the omentum and small intestines may be found attached.

Diagnosis.—There are a history and evidence of infection. A suppurating ovarian cyst is usually unilateral and if of large size pins the uterus up against the pubes. An ectopic gestation is also on one side, is harder than a pyosalpinx, and gives subjective symptoms presumptive of its existence, such as irregular bleedings, stabbing pains, attacks of syncope, etc. A broad ligament cyst or abscess is always sessile upon the uterus, giving no sulcus between the cyst and the uterus, and is always situated laterally. Broad ligament fibroids are of cartilaginous hardness and are also sessile upon the uterus. All broad ligament growths are part of the uterus so far as mobility is concerned. Exploratory puncture is not a safe procedure, inasmuch as very often other abdominal contents lie between the tube and the vaginal vault. Besides, failure to find the pus upon puncture is no proof of its absence. Infected hematoma following ruptured ectopic gestation, where the clot is in the folds of the broad ligament assumes the characteristics of other fluid accumulations in the ligament. Blood clotted and free in the pelvis, infected and encapsulated by lymph, forces the uterus forward and immovably fixes it. The lateral fornices are symmetrically occupied by the clot. In some cases a positive diagnosis cannot be made without a posterior vaginal incision, a perfectly safe procedure.

Treatment.—Pus in a preformed sac (pyosalpinx) is radically cured only upon extirpation of the sac. Incision and drainage will relieve, but not insure against a relapse at some distant day. Inasmuch as a pus tube is rarely found upon one side and normal adnexa upon the other, vaginal ablation of uterus and adnexa is the indicated radical operation in all cases. If ablation is not accepted the preferable operation in all cases of pyosalpinx purely pelvic in their relations and important associations, is evacuation through the vagina by broad incision. This treatment is to be practiced to the exclusion of laparotomy and removal in all except a limited number of cases. The excepted cases are the few in which appendicitis coexists with a pyosalpinx, and cases where a fistula exists between the *small* gut and tube. I have never seen a gonorrheal pus tube upon one side, in which there was not sufficient tubal or ovarian disease upon the other to warrant its removal if it existed alone. The treatment is either wholly conservative (see Conservative Treatment) or thoroughly radical.

Sequelæ.—Very rarely these pus tubes rupture into the general peritoneal cavity. But in many cases they leak slowly, causing the effusion of large masses of isolating lymph, beneath the plates of which the pus oozes (see Diffuse Pelvic Suppuration). Pus tubes form adhesions with the intestines, usually with the sigmoid, and may rupture into it. If the opening is into the sigmoid the pus is discharged *per rectum* and the pus-sac becomes additionally infected from the gut. If the fistula is into a knuckle of small gut the pus will be absorbed. The

SALPINGITIS.

pus may also find its way into the vagina or even into the bladder. The fistulæ are not permanent, but open and close in obedience to the distention of the pus-sac. Pyosalpinx upon the right side is very frequently associated with, and adherent to, a diseased appendix vermiformis. Life is destroyed by the prolonged suppuration, and nephritis is a common sequela. In questioning over three hundred cases of phthisis, I found that four in every seven never had any lung symptoms until the onset of an attack of pelvic inflammation. The general debility

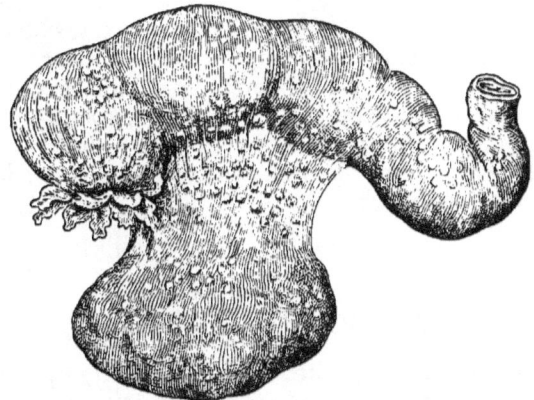

FIG. 27.—Tuberculosis of the Fallopian tubes. The disease has extended to the peritoneum, which is covered with tubercles (Penrose).

following the latter conduced to the inception of the former. Not pelvic inflammation but nephritis and phthisis end the lives of most of these sufferers.

TUBERCULAR SALPINGITIS.

This is always chronic; and it is doubtful if it arises primarily, being secondary to tubercular peritoneal disease. I have seen one case in which I thought it primary. The disease may be due to a general tuberculosis, to extension from a tubercular intestinal ulcer, or be set up by tubercle bacilli introduced through the uterus by means of dirty instruments (Fig. 27). The tubercles lie upon the peritoneal surface of the tube, as well as in the cavity, forming miliary tubercular salpingitis. Or the disease may assume either of the two other common tubercular types: caseous infiltration, or chronic fibroid tuberculosis. In one instance, there will be a tube filled with cheesy pus and studded with tubercles; while, in another, the tubercles are few, and the production of connective tissue marked. Tubercular disease constitutes about fifteen per cent. of all inflammatory pelvic disease of a chronic type.

The symptoms and signs are those of pyosalpinx, or other chronic tubal disease, due to other causes.

Treatment.—Whenever upon exploratory vaginal section tubercular adnexal disease is found, ablation should be performed. But this statement may be qualified somewhat by excluding operation when general peritoneal tuberculosis coexists.

PELVIC PERITONITIS.

The normal pelvic peritoneum is generally transparent, and through it the color of the underlying tissue may be detected. In certain portions of the pelvis it is thicker than at other points, notably over the rectum and the iliac vessels, and at these points the peritoneum is opaque. Over the uterus the peritoneum is thin, and the peritoneal covering of the tubes is exceedingly delicate.

When inflamed the peritoneum becomes deeply injected. Its color will vary from a delicate pink to a livid hue, according to the severity of the process. At first serum is poured out in a variable quantity. As the circulatory stasis increases the endothelial cells shrink away from each other, and the underlying lymph spaces are exposed. White blood-corpuscles and plasma cells pass out upon the surface of the membrane, where they form masses of "lymph." If the process subsides, these lymph masses change into connective tissue bundles or "adhesions," which become supplied with blood-vessels and are covered with endothelial cells. If the infecting agent overwhelms the vital forces, the cells die and produce pus. According to the nature of the result of the infection, we have either a slightly injected peritoneum with serum as a result, or one deeply colored, smooth, and shining generally, but at points unglazed, and covered by lymph; or one livid in hue, rough in appearance, and devoid of endothelium, studded here and there by small lymph masses, and showing frequent spots of purulent lymph. Pus in the pelvic peritoneal cavity may be whitish, yellow, or greenish-yellow. Usually it is odorless, but it may be tainted with intestinal gases without there being an opening into the gut.

Causes.—The causes of pelvic peritonitis may be classified as *direct* and *contributing.*

DIRECT.—Pelvic peritonitis in women is caused by *colon bacilli, gonococci, staphylococci, streptococci, tubercle bacilli,* and more rarely by other pathogenic germs. A certain form of peritonitis is also produced by the chemical irritants which are contained in antiseptic dressings, when these touch the peritoneum. According to the nature of the infection the character of the lesions will vary. Where *colon bacilli* cause the inflammation, there is but little lymph produced, and not much of serum; but it is doubtful if suppurative peritonitis is ever set up by the colon bacillus alone. Peritonitis caused by the colon bacillus is less active than any other, and the local disturbances are slight. It is found most commonly as a result of inflammation of the colon when this is accompanied by bowel distention and retention of feces. Clinically, we meet with it most often after it has produced adhesions, or in its acute stage as a complication in the after-treatment of intrapelvic operations. Very slight toxemia is produced by it, and hence, the rise in temperature and pulse rate may be so slight as to be unnoticed. In cases of adherent retroposed uteri which we meet with in young women who have never had uterine or tubal inflammation, the adhesions are probably always due to the colon bacillus. Whenever any organ rests immovable upon one spot of the large gut for a length of time, migration of the colon bacillus is apt to result, causing limited effusion of lymph, and the ultimate formation of delicate adhesions. In the mistaken treatment of abdominal diseases by opiates, the migration of colon bacilli is facilitated.

Where peritonitis is caused by *gonococci*, serum and lymph are produced. Under ordinary conditions gonococci do not produce purulent peritonitis. Suddenly flooding the pelvic cavity by a large quantity of virulent gonorrheal pus will set up a purulent peritonitis. Pelvic peritonitis due to gonorrhea may be caused by the gonococci reaching the peritoneum through the uterus and Fallopian tubes, through the bladder, or through a ureter. By far the greater number of cases of this form of peritonitis are produced by the infection coming through the uterus and tubes. Gonorrhea causes peritonitis by extending directly along the continuity of the tissue, and not through the medium of the lymphatics. As a consequence we have the peritonitis secondary to a salpingitis. The first effusion of plastic lymph occurs at the fimbriated end of the affected tube, effectually closing it, and causing it to unite to any adjacent organ. As the infection progresses, lymph is thrown out upon the surface of the uterus as well as on the tube. The production of serum is slight, and altogether the tendency of the peritoneal inflammation is to localization. A characteristic of gonorrheal peritonitis is its tendency to recurrence. The younger the subject infected, the more pronounced the peritonitis. The puerperal state after the third month appears to grant a certain immunity against this form of infection. It is generally seen in unimpregnated women. For the lesions induced in the ovaries and tubes by this infection the reader is referred to "Salpingitis" and "Ovaritis."

Pelvic peritonitis due to *staphylococci* usually results in the limited production of both serum and lymph. I do not believe that primary purulent peritonitis is ever caused by staphylococci. The infection may reach the peritoneum through either the medium of the tubes or through the lymphatics, or by both channels. In the former case the peritoneum at the fimbriated end of the tube is first affected, resulting in the closure of the tube. Here the pelvic peritonitis is limited. Where the infection passes to the peritoneum through the lymphatics, the peritonitis occurs as a primary disease. According to the extent of the infection the severity of the peritonitis will vary. The passage of an unclean sound which bruises a slight area of endometrium will cause but a limited degree of infection and a small amount of lymph will be poured out upon the broad ligament or uterine wall. The same kind of infection occurring at the site of a recently detached placenta may result in an infection which will be the cause of a general plastic pelvic peritonitis. The degree of the peritonitis will correspond to the number of lymphatics which are involved. If this infection occurs in the puerperal state, and results in the production of a broad ligament phlegmon, this may be the means of starting a suppurative type of peritonitis. But this suppurative peritonitis will be secondary to a plastic form, and the pus will be locked in. Diffuse suppuration in the pelvis will result, but not primary purulent peritonitis. There is a vast difference, both in the local lesions and danger to life. The tendency of infection by the colon bacillus is to produce lymph-effusion at the point of migration of the bacillus, and this will result in attaching that point to any organ which rests upon it. Primary tubal and ovarian diseases are not produced by the colon bacillus.

The tendency of infection by the gonococcus is to produce primary suppurative salpingitis with secondary peritonitis about the tubal orifices. This focus of suppuration in the tube becomes the agent by which other attacks of peritonitis are produced. Recurrent plastic peritonitis is characteristic of gonorrheal infection. The tubal lesions are more marked than those of the peritoneum.

Staphylococcus infection tends to produce: (*a*) tubal inflammation with secondary peritonitis; and, (*b*) pelvic lymphangitis with primary peritonitis. The extent of the peritonitis is greater than where the gonococcus is the infecting agent.

None of these pathogenic germs tends ordinarily to produce primary purulent peritonitis. They are usually local in their activity and produce but mild toxemia.

Such is not the tendency of the *streptococcus*. From its first introduction into the system this germ produces the greatest amount of septicemia relative to the degree of the local disturbance. Introduced into an unimpregnated uterus, it produces either tubal inflammation with peritonitis secondary to that, or primary peritonitis by extension through the lymphatics. Occurring in the absence of a recent gestation it results in the liberal outpouring of serum and the widest effusion of lymph by the peritoneum. As a result, there is suppuration in the tube or ovary, or both, which is surrounded by large masses of lymph. There is never intermission, however, in an inflammation produced by the streptococcus. It never becomes strictly a local disease, but there is a continuation of acute manifestations with marked exacerbations. After a time the pus leaks into the lymph planes and the gravest form of diffuse pelvic suppuration is produced. The streptococcus is found in the products of inflammation in such pelves, but there are such marked differences in the gravity of the symptoms in various cases, that we are forced to believe that there is a great variability in the virulence of this germ.

Occurring in a uterus recently aborted or delivered, this form of infection may result in *primary purulent peritonitis*. This is the gravest form of peritoneal inflammation. Large quantities of serum are produced; the peritoneum is livid in color; the effusion of plastic lymph is limited, and as a result, there is little or no tendency to localization of the disease. Death may occur before any lymph is effused upon the pelvic peritoneum, there being induced a complete stasis in the local vital forces. From the first initiative chill to death may be less than three days.

If, however, the septicemia is not so rapidly fatal, the infection tends to cause death of the cells thrown out by the peritoneum. The pelvic peritoneum becomes granular in appearance, and isolated spots of pus and lymph are scattered over its surface. There is slight tendency to union between the organs in contact. The lower pelvis is filled with fluid, sometimes straw-colored, with isolated pus cells and flocculi of lymph, or cream-colored from admixture of pus. The fluid has a putrid odor in many cases. The least touch suffices to break the peritoneum, which is denuded of its endothelium. The tendency of purulent pelvic peritonitis is to become purulent general peritonitis. It is always fatal, unless checked by operation, causing death either by heart paralysis from the effects of its toxins, or by producing endocarditis, pneumonia or nephritis.

Sometimes, usually in cases not aborted, or in those aborted before the third month, the peritoneum will be able to resist the streptococci sufficiently to produce large masses of vitalized lymph. If it does this, the pus which the infection has produced in the broad ligament or tube or ovary will be locked in. If this localization of the suppuration is once obtained, the peritoneum throws out enormous masses of lymph. The intestines, omentum, bladder and fundus uteri become firmly united at the pelvic brim, and the pelvic suppuration is effectively locked in.

Streptococcus infection, if not fatal in a few days, always results in suppuration somewhere. Occurring in the unimpregnated uterus, the pus may be produced in tube, ovary or broad ligament, with diffuse plastic peritonitis as a complication. Occurring free in the pelvic cavity, the pus is located according to the posture of the patient, as any other fluid would be.

CONTRIBUTING CAUSES.—Although we find that all

forms of pelvic peritonitis are due to some irritant, either of a germ or chemical nature, these direct causes are not always operative unless there exist local conditions which are propitious to their activity. The peritonitis due to gonorrhea we find more often occurring at the *menstrual periods*. At this time the physiological exfoliation of the epithelium of the endometrium conduces to the introduction of the gonococcus into the pelvis. Furthermore, gonorrhea may for years remain latent in the cervical canal, and a general gonorrheal infection of the pelvis be set up by operations upon a uterus so infected. We may consider *trauma* as a contributing cause. Any injury to the general system which will produce marked stasis in the pelvic circulation will also suffice to bring on an attack of peritonitis where latent pathogenic cocci are present. Such an agency may be *prolonged exposure* to cold, and great physical effort. Those germs which are common in the bowel operate as causes of peritonitis when the bowels are inactive and feces are retained. *Chronic constipation* undoubtedly conduces to peritonitis of one form. The breaches of surface incident to *abortion* and *labor* are particularly conducive to the onset of an attack of peritonitis, by furnishing points for the entrance of germs into the system.

Symptoms.—As pelvic peritonitis produces three kinds of effusion, serum, lymph, (resulting in adhesions recent or old), and pus, the local signs will vary greatly. The degree of toxemia will be largely governed by the nature of the infecting agent, and there is a wide range in symptoms. The peritonitis which accompanies gonorrheal and septic infections is so commonly associated with or caused by tubal diseases that the reader is referred to the chapter on "Salpingitis" for its description.

There is undoubtedly a period of incubation from the introduction of pathogenic germs to the first evidence of peritonitis. Just how long this is we do not know, but I have thought that where the infection travels through the tubes it more rapidly produces peritonitis than where the lymphatics are the carriers. To this, however, I must make the exception that in infection occurring after the third month of gestation the peritonitis usually occurs directly as a result of the lymphatic infection, and not through the medium of the tubes. In any case, from two to three days elapse from the time the germ of inflammation is introduced into the uterus to the first evidence of peritonitis; and a few hours only are needed for the peritoneum to develop some evidence of inflammation after being brought into contact with germs.

There is usually at first a free effusion of serum. This I have never been able to determine upon examination, and have found it only after the peritoneum has been severed.

When lymph is first thrown out it is colorless. Within a few hours vessels appear in it, and it becomes organized into a new tissue. Lymph tends to hold immobile the organs between which it lies. As a consequence we have fixity. If the lymph be exuded upon the broad ligaments only, these are thereby stiffened, and bilateral mobility of the uterus becomes limited. Furthermore, the thickened broad ligaments have lost their elasticity, and the examining finger finds a density in each lateral vaginal fornix, where formerly there was perfect elasticity. The uterus is fixed, and at the sides of the cervix are dense masses of exudate. In extreme cases of plastic effusion the uterus will be as immovable as though resting in the knot-hole of a board, and the density, inelasticity and infiltration will be all around the uterus. If the lymph be effused upon one broad ligament only, the cervix can not be moved away from the affected side, and in drawing down the uterus the cervix will swing toward the firmer ligament,—it will not come down in the middle line of the vagina. Masses of recent lymph are soft, but do not give out fluctuation. The sense of a mass is produced as much by the edema of the tissues as by the lymph. Where the lymph has been effused about a tumor or pus focus, it increases the bulk of the mass and fixes it. And when the outpouring of lymph has been repeated, the density of fluid accumulations is increased, leading to their being mistaken for solid tumors. The effect of thick lymph accumulations upon involved organs is of interest. If the effusion has taken place about a bladder which has been neglected in over distention, that organ will be fixed high up and cannot be completely emptied And if distention of the bladder is prevented by oft-repeated catheterization while the lymph is being poured out, the bladder will be fixed in systole, and distention will be impossible so long as the adhesions remain. As I have pointed out, lymph tends to fixation of the uterus. The same is true of the ovaries and tubes where they are implicated in the effusion; they remain attached to whatever organs they may rest against at the time the lymph was effused about them, and can move only as those organs move. Where the lymph is secreted about a rectum distended by feces or gas, the gut remains canalized. As a consequence we have a distended rectum, one whose walls are never collapsed, a common feature of diffuse pelvic peritonitis. If the rectum be empty when large masses of lymph are produced about it, it will be partially strictured. The ureters pass beneath both broad ligaments under the peritoneum. Recent effusion of lymph has little effect upon them. If a knuckle of small gut is caught and fastened by lymph in the pelvis, its function is markedly interfered with, chiefly in the matter of rhythmical peristalsis. The adhesions which result from lymph, while much less than would be expected to follow so generous a production of this material, yet produce grave consequences. The bladder may be held at the fundus uteri, permanently distended. The retroverted uterus may be fixed to the rectum and be capable of replacement only by ballooning out the rectum. The contraction of the lymph upon the broad ligament causes stricture of the ureter and hydro-ureter. The tubes are distorted and strictured, the pelvic vessels obstructed, and a condition of atrophy in the genital organs results from impeded circulation. The adhesions to the small intestines are continuously pulled against. As a result, they are teazed out so as to be many inches in length, forming bridges across which loops of intestine may fall and become strangulated. Adhesions between the adnexa and the peritoneum over the psoas muscles may be so stout as seriously to impede bodily movements. An interesting tendency of these false unions is that their vessels furnish additional nourishment to the attached organs. An ovary may become detached from its normal site and be entirely supported through false bands. The appendix vermiformis may receive its sole blood supply through a new attachment to the right appendages, and slough off when these are severed. Neoplasms have been found separated from the uterus and nourished through adhesions. The spleen, if attached in this way, becomes enlarged. Omental adhesions often produce at the seat of attachment large masses of fat tumor through the influence of the vessels running through the new bands.

Where intraperitoneal pus is the result of peritonitis, the pelvic organs are all fixed, partly by lymph and partly by subserous edema. There is marked lack of elasticity in the vaginal vault. If the pus be very fluid it cannot be detected, but when thick a spongy bulging may be felt in the posterior vaginal fornix. This changes with change in the posture of the patient, and, unlike very recent lymph, can be displaced upward by pressure. Sensitiveness to pressure is slight when serum alone is produced; where lymph is effused it is marked; and in primary suppurative peritonitis the lack of pelvic sensitiveness and pain is a marked feature. In fact, in the worst cases of purulent peritonitis there is pelvic analgesia. This is an important sign.

TYMPANITES.—This is noticeable in cases of purulent and lymph effusion, but is not present in serous effusion. It is dependent upon the degree and kind of infection, the state of the bowels, and the *medication*. If the bowels are kept empty by washings and no opiates given, the tympanites is not marked in pelvic peritonitis.

SUBJECTIVE SYMPTOMS.—*Pain* is slight where serum alone is produced and when the infection is followed by very gradual effusion of lymph. When intraperitoneal pus (purulent peritonitis) is present, pain is not marked and may be entirely absent. Where a sudden sharp outpouring of lymph takes place, the pain is severe and continuous. It is exceedingly difficult to say how much pain is the result of the involvement of the peritoneum, and how much is caused by coincident inflammation of uterus, ovaries or tubes. Movement of the organs over which lymph is effused increases the pain. Sudden pelvic pain, accompanied by grave general symptoms, pyrexia, and septicemia, and followed by a rather abrupt cessation of the pain, points to purulent peritonitis.

Temperature.—Where serum alone is produced, the temperature is seldom elevated one degree. (I consider $99\frac{2}{5}°$ as normal rectal temperature.) The height of the fever accompanying lymph effusion will vary with the patient's general condition, the kind of infection present, and the degree of the infection. Gonorrheal peritonitis rarely produces a temperature above $102\frac{1}{2}°$. The same is true with staphylococcus infection. In streptococcus poisoning the temperature rapidly reaches $103°$, and is more often above than below that point. The evening temperature is generally a degree above the morning. A temperature holding steadily for more than a day above $103°$ should create great uneasiness in the medical attendant. This is particularly necessary when the fever comes on after operation, abortion, or labor. There are marked fluctuations in the temperature in most cases. The falls in temperature will be found to correspond pretty accurately with an increased excretion of urine and evacuation of the bowels. Through the kidneys and the bowels toxins are eliminated.

The *pulse* in gonorrheal peritonitis seldom reaches $110°$. Where the infecting agent is the streptococcus, the pulse rarely falls below $110°$ beats a minute. Of more value in determining the nature of the infecting agent than either pulse or temperature alone, are their relative marks. Thus, a temperature of $103°$ with a pulse less than $110°$ need cause little apprehension as to the ultimate result, while the same temperature with a pulse of 130° calls for immediate interference, and is indicative of a virulent infection, probably streptococcic. The effusion of lymph does not cause the rapid pulse and fever. Great masses of lymph may be thrown out about a gauze drain in the pelvis, and yet the pulse be but slightly accelerated and fever be absent. The fever and quick pulse accompanying those degrees of infection which result in lymph effusion are produced, not by the lymph, but by the toxins of the invading germs. We find slight rise in temperature attending the production of large plaques of lymph, and high temperature where no lymph is produced. The fever is due to the toxemia, not to the outpouring of lymph

Rigors.—Chills are not features of peritonitis, except when there is a sudden rise in temperature from a point near 101° to one 4° or 5° higher. Then a slight rigor will mark the inception of the rise in many cases. Rigors may be stated to be features rather of a general septicemia than of a localized peritonitis.

Digestive Symptoms.—There being an overproduction of bile, vomiting is apt to follow overloading the stomach in cases of pelvic peritonitis. Beyond this, vomiting is rarely present as a symptom of pelvic peritonitis. The onset of persistent vomiting, where not produced by improper food and drugs, if accompanied by high pulse and temperature, is alarming. It is indicative of a peritonitis which is extending above the pelvic brim. The bowels are prone to costiveness in pelvic peritonitis, on account of interference with their peristalsis by adhesions, and in part to the increase in pain produced by defecation. In purulent peritonitis there is, on the contrary, very commonly a diarrhea.

The kidneys are rarely affected in any forms of infection save one. The urine is increased in amount and the percentage of urea is increased. In streptococcus infection acute parenchymatous nephritis is a common complication.

The heart and *the lungs* are not affected in any form of

peritonitis, except that due to streptococcus, the purulent form. Endocarditis, pneumonitis, and pleuritis are very often met with in cases of purulent peritonitis. It is rare for a case of streptococcus infection to recover without some grave complication.

Diagnosis.—The diagnosis of pelvic peritonitis is generally embraced in that of some one of its accompanying lesions, salpingitis endometritis, etc. It is not so difficult to detect an effusion of lymph in the pelvis. I have never been able to determine the presence of the serum which I have evacuated so many times. The important and difficult task is to diagnosticate the presence of primary purulent pelvic peritonitis. I may mention the more usual features of this disease: usually a history of criminal abortion or instrumental labor, often an initiative chill, pulse from the first 110° or more, temperature at or above 103°, tympanites, not much pain, stupid face, tendency to somnolence, sordes on teeth, red furred tongue, muttering delirium; uterus fixed in pelvis, vaginal vault hardened, spongy mass in posterior cul-de-sac, not much sensitiveness. The woman *looks* very ill in a a day. It is especially difficult to differentiate suppurative pelvic peritonitis from general suppurative peritonitis. I have never found a case of suppurative pelvic peritonitis in which there was not a history of either abortion, labor or trauma. In general suppurative peritonitis there is no such history; it is usually due to appendicitis. It is impossible to determine just when a suppurative peritonitis arising in the pelvis ceases to be pelvic and becomes general.

Prognosis.—Where the effusion is purely of serum and lymph there is no risk to life. All cases of purulent peritonitis die unless operated upon, and most of these perish. In early surgical interference lies the only hope of saving the lives of these women.

If lymph effusions are allowed to remain they produce permanent lesions.

When a woman has once had pelvic peritonitis with the production of lymph she has before her all her life the possibility of an operation of some sort. The prognosis of peritonitis is markedly influenced by the nature of the causative lesion, whether gonorrheal endometritis, salpingitis, pelvic lymphangitis, etc. These are discussed elsewhere.

Treatment.—If an effusion of serum alone is suspected the treatment consists in preventing further extension of the process by removing the causative focus of infection. The serum will then be absorbed. If lymph is effused, the cul-de-sac should be opened, and all attachments between the viscera should be severed, after the focus of infection (usually the uterus) has been cleansed. (See Exploratory Vaginal Section). This is necessary because the lymph in contracting into bands produces such distortion of the viscera as will destroy their function, partially at least. It is to prevent tubal and ovarian suppuration, as well as future adhesions, that this operation is recommended. (See Salpingitis.)

SUPPURATIVE PELVIC PERITONITIS demands the most energetic measures. The operation to be applied is purely an evacuative one. In most cases it will suffice to open the posterior cul-de-sac, let out the pus and fill the pelvis with iodoform gauze. But in all cases of purulent peritonitis, the Mikulicz dressing of iodoform gauze is absolutely necessary to remove the large quantities of septic fluid which escape after the operation, and to furnish iodine in the form of iodoform in order that the streptococci may be destroyed. The author has shown that this result follows the use of a certain form of gauze in these cases.

I have had no experience with the use of antistreptococcus serum in these cases, and cannot see how it can benefit them *before* an operation. Given after an operation it may prevent those complications which commonly cause death. Certainly, preliminary reports warrant its trial, but not to the exclusion of measures here recommended.

GENERAL TREATMENT.—When a heart stimulant is

needed in pelvic peritonitis I employ strychnin. As a rule this will be found necessary in the purulent type only. Here large doses must be administered, beginning with gr. $\frac{1}{30}$ q. 4. h. hypodermically and gradually increasing. I dislike to give alcohol except in the form of champagne. If strong liquors are given, brandy in six parts of iced water is best. Of brandy ℥ss. q. 3. h. is an average dose. Again this is needed in the purulent type only. For local pain, blood-letting from the cervix and ichthyol tampons 10 per cent. furnish greatest relief. The colon should be washed out daily with a quart of normal salt-solution. In most cases I allow half diet; but in the purulent type I employ an exclusive liquid diet— one ounce of beef juice every four hours and two ounces of chicken broth every four hours. These are made to alternate, at two-hour intervals. Between feedings an abundance of water, with a few drops of lemon juice to acidulate it, is given if vomiting is present.

The general treatment of pelvic peritonitis should be sustaining. Inasmuch as the effusion of lymph serves a good purpose at first, it is to be interfered with only *after* the causative focus of infection is cleansed. Then, for evident reasons, the effusion of lymph must be checked. If pus is present, it must be evacuated so soon as discovered.

TUBERCULAR PELVIC PERITONITIS.

I have reserved a description of tubercular pelvic peritonitis for a separate section, since its lesions differ in character from those produced by pathogenic cocci. The bacilli reach the peritoneum through either the blood or the lymphatics. It is of interest that tubercular peritonitis is not a result of a similar process in the uterus, but that the peritonitis produced through the blood is merely part of a general tuberculosis, and that which extends through the lymphatics is from some tuberculous focus in the abdomen. The origin of certain cases, however, is obscure.

When the tubercles appear beneath the peritoneum, they are in the form of gray nodules. The peritoneum is at first unchanged. Soon serum is effused. The peritoneum becomes congested, and the endothelium becomes multiplied and exfoliates. Leukocytes escape from the vessels, and blood may tinge the serous fluid. The process may stop here, and recovery may take place. Or, lymph may result from the presence of the tubercles and intervisceral adhesions result. The tubal orifices may close, and retention cysts be formed. The tubercles tend to invade the tubes and produce tubercular salpingitis. Where suppuration results the pus-producing agent is not the tubercle bacillus, but some complicating pyogenic coccus.

The symptoms are those of pelvic peritonitis due to other causes that produce serum and lymph. Neither by symptoms nor examination can the disease be differentiated from other forms. It may be suspected when peritonitis occurs slowly, without marked acute symptoms, without evidences of primary uterine and tubal disease, and when great emaciation and debility appear without assignable cause.

The treatment is always to be evacuative. The uterus need not be curetted. The cul-de-sac is opened and all adhesions severed. After this the pelvis is irrigated with normal salt-solution and a high Mikulicz dressing is applied. The strips of gauze should extend quite to the fundus. If secondary tubal and ovarian disease are present, vaginal ablation is indicated. The exposure to air, the trauma incident to the operation, and the iodine in the gauze probably effect the cure.

Serous pelvic peritonitis is innocent; plastic lymph-producing peritonitis is beneficent, but purulent peritonitis is the most fatal of all diseases affecting the human body.

Peritonitis is not to be considered in the light of the

results of the process, but rather in a knowledge of its causes.

The greater the lymph effusion the less immediate danger to the patient, for the lymph tends to lock in the infection and limit it, and in that sense it is beneficent. But in so considering it we must not be consoled into an ignorance of the ultimate results of a generous outpouring of lymph; and it is our duty to check it while checking the infection.

INFLAMMATORY DISEASES OF THE OVARIES.

Acute Peri-ovaritis.—The infection may extend to the peritoneal covering of the ovary from the tube; or the ovarian peritoneum may become inflamed conjointly with the adjacent pelvic peritoneum from an infection which has reached it through the lymph streams. In other words, pelvic peritonitis from any cause may implicate the peritoneum of the ovary. The type of inflammation is the same here as in other portions of the pelvic peritoneum. There is an effusion of lymph which causes the ovary to become attached to adjacent organs, most commonly to the tube and to the broad ligament. The entire ovary enlarges and appears edematous. Upon its surface flakes of lymph are seen, or the entire organ may be covered by a thick plastic deposit. If the process subsides, there result delicate false bands attaching the ovary to some portion of the pelvic contents, or the union may be so broad that the ovary is firmly plastered to the uterus, broad ligament, tube, or lateral pelvic wall. United in this way, and repeated attacks of peritonitis occurring, the ovary may be entirely isolated from the general pelvic cavity and lie in a pocket formed by sheets of new membrane. In all cases the capsule of the ovary is thickened. Periovaritis may extend to the stroma of the ovary and to the follicles. The stroma may become infiltrated with new connective tissue elements and on contracting produce "ovarian sclerosis" (see Fig. 21). If the follicles are involved, they become enlarged and, unable to discharge their contents through the thickened capsule, they present the characteristics of permanent cysts. The ovary is "cystic" and enlarged. Between the cysts are found areas of sclerosed tissue (Fig. 28). The cysts project beneath the capsule and appear as pearl-like bodies. Upon evacuating one a thin serous or tenacious glairy fluid escapes, and the cyst wall collapses. The cysts may com-

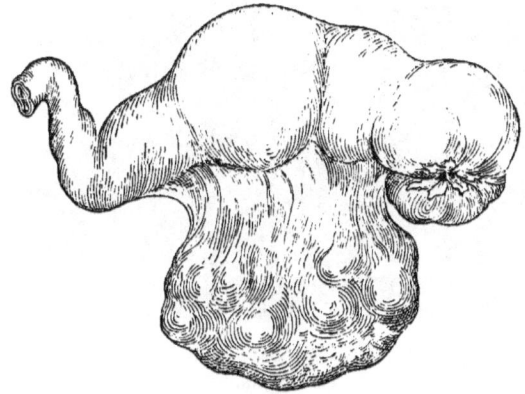

FIG. 28.—Salpingitis with partial inversion of the fimbriæ. Cystic degeneration of the ovary (Penrose).

municate with each other and large cavities be thus formed. Cysticdegeneration must not be confounded with ovarian cystoma (ovarian cyst). They are essentially different. Ovules are found in the cysts of cystic ovaritis and women with such ovaries conceive. We may therefore consider this lesion unimportant, and inasmuch as the organs so affected functionate they should be preserved. Blood may be extravasated into one or more of the cysts, constituting "ovarian apoplexy." The walls of the blood cavity are lined by a membrane, sometimes dark, in other cases yellow, which is loosely attached to the surrounding ovarian stroma. There may be but one such blood cyst, often several inches in diameter, or there may be a number of small ones.

The lymphatics of the ovary may be chiefly affected and the ovary become soft, edematous and much enlarged, even four times its normal size. Upon splitting such an ovary it appears gelatinous. This is "edematous ovaritis" (Fig. 29). The kind of infection brought to the ovary by the lymph streams may be so virulent that suppuration takes place in the stroma; or an acutely inflamed

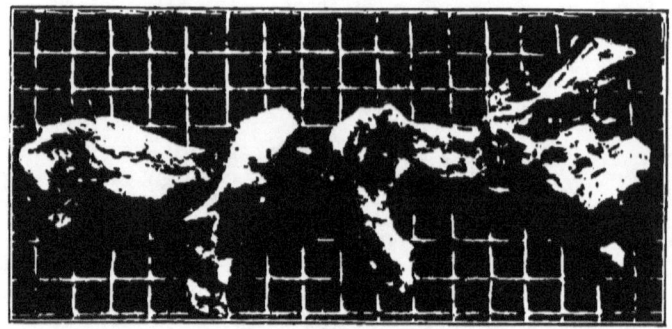

FIG. 29.—Bilateral pyosalpinx, left edematous ovaritis, and subserous cyst. Vaginal ablation.

tube may become sealed to the ovary and the ovary suppurate from proximity to the pus tube. "Ovarian abscess" results (Fig. 30). These pus ovaries are always enlarged. Sometimes there are a great number of small foci of pus; in other instances the ovarian capsule surrounds one large pus cavity. I have removed by the vagina one which lifted the uterus up out of the pelvis and completely filled the latter.

Surgically the inflammatory states of the ovary may be divided into non-purulent and purulent. If pus is not present, attempts should always be made to save at least portions of the organ. Sclerosis of the ovaries may coexist with a like process in the tubes and the uterus. Manifestly it is useless to preserve the diseased ovaries

INFLAMMATORY DISEASES OF THE OVARIES. 109

when there exists an indication for removing the uterus and tubes in such a case.

Symptoms.—It is the author's belief that non-purulent inflammatory disease of the ovaries produces few symp-

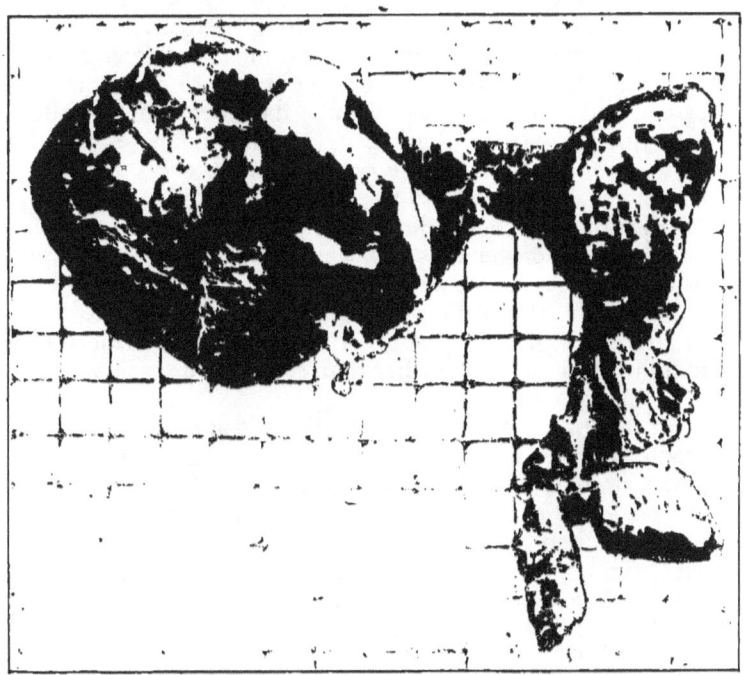

FIG. 30.—Very large ovarian abscess, and half the uterus. Vaginal ablation.

toms. It is not the cystic or apoplectic ovary which causes the distress, but the co-existing tubal and peritoneal disease that is commonly found in these cases. Such being my view, I am disposed to apply conservative treatment to cystic and apoplectic ovaries.

In the young girl the ovary is pink and has a delicate capsule. After hundreds of ova have torn their way through the capsule, it becomes scarred and pale; the capsule is thickened, and the ovary distorted. No two ovaries are exactly alike; some are round, some long, some of hourglass shape; some measure a half inch in length, others as much as two inches. In other words, there is the greatest variety among perfectly normal ovaries. But surgeons have spayed thousands of women because their ovaries did not conform to some ideal organ, and many of these women have been sent to the mad-house. Hystero-epilepsy, epilepsy, neuroses of all sorts, chronic pelvic pain, in short almost every obscure complaint in women, has been treated by the removal of ovaries that were cystic, apoplectic, or "atrophied." I believe that non-purulent ovaries produce few symptoms other than a sense of weight when they are large. When adherent in the cul-de-sac and compressed by other organs, they give pain; but it is the lack of freedom in mobility and position, rather than essential disease, which is to blame.

Periovaritis gives no distinguishing symptoms, inasmuch as it is always accompanied by some more important lesion, as salpingitis or pelvic peritonitis.

Ovarian Abscess can not be differentiated from pyosalpinx. The history will sometimes presumptively indicate the character of the abscess. Ovarian abscess is usually due to infection after abortion or labor, and when due to gonorrhea, it is found as a lesion secondary to salpingitis (Fig. 31). The symptoms are the same as those of pyosalpinx. Upon examining a pus-ovary case we do not get fluctuation. A firmly adherent, dense, sensitive mass is found to one side or behind the uterus. There are evidences of acute pelvic peritonitis, fever, pain, etc., just as are found with pyosalpinx. Still I have seen a case of an enormous pus-ovary holding a pint in which there was absolutely no evidence of fever.

Treatment.—*Acute Periovaritis.*—Inasmuch as this condition is not found existing alone, but as a concomitant of inflammation of other portions of the pelvic peritoneum, there is no special treatment to be directed to it.

INFLAMMATORY DISEASES OF THE OVARIES.

Blisters and iodin applied to the abdomen over the ovaries are classical, but are of doubtful efficacy. I have found that the maintenance of a definite warmth over the abdomen by employing moist dressings which are covered by rubber tissue, painting the vault of the vagina with 10 per cent. to 20 per cent. ichthyol in boroglyceride, and keeping the bowels washed out so that hard fecal masses do not press on the ovaries, afford the greatest relief. If this condition is found to exist after a cul-de-sac operation is made, the ovaries should be detached from their false attachments.

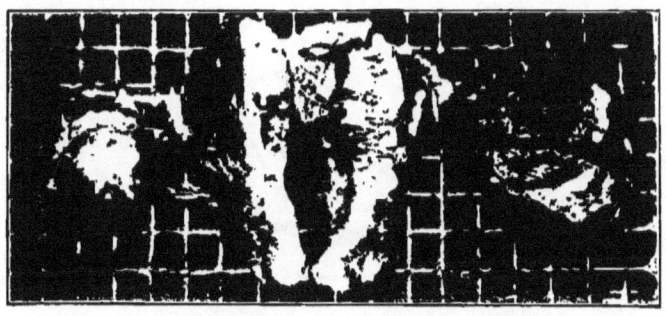

FIG. 31.—Right pyosalpinx and ovarian abscess. Left ruptured ectopic gestation. Left ovarian apoplexy. Vaginal ablation.

Ovarian sclerosis cannot be cured by any means. Such ovaries may be removed when indications exist for removing the uterus, but sclerosis of the ovaries only does not warrant their removal.

Edematous ovaritis I have not met with except under circumstances which required removal of all the generative organs.

Cystic Ovaries.—Upon opening the cul-de-sac I first attempt to free the ovary of one side. When this is loose I introduce a posterior retractor into the pelvis, and with the trowel lift the uterus into the abdomen. A gauze pad is next introduced between the retractors, and the head of the table is lowered. If the intestines are not adherent they will escape into the abdomen. The ovary is now grasped with Luer's forceps and pulled down. A pair of stout mouse-tooth forceps or bullet forceps may be substituted for Luer's instrument. The surface of the ovary is inspected carefully, and all cysts are stabbed with a tenotomy knife. The bleeding is trivial. When all the cysts are evacuated, the ovary is returned to the pelvis and the other ovary similarly treated. I am opposed to igni-puncture with the Paquelin cautery. This method of evacuation is uselessly complicated, and the healing after it is not normal. After returning the ovaries, the pelvis is wiped dry, and the gauze pad is

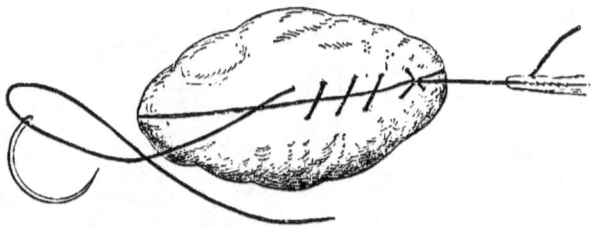

FIG. 32.—Suture of resection wound in the ovary.

removed. The rent in the cul-de-sac is sewed up, if the uterus is not retroposed, if the patient has not purulent endometritis, and if there is not pronounced oozing in the pelvis. When either of these exists it is better to introduce a plug of iodoform gauze into the opening and pack the vagina.

Ovarian Apoplexy.—Having released the ovary from false attachments, it is pulled into the vagina. In doing this care is exercised and the forceps should grasp the more normal portions of the organ. Steadying the ovary the surgeon splits the periphery of the blood-sac with scissors. Fluid and old blood escape and should be caught with gauze. Holding apart the lips of the rent, the lining cavity of the sac is easily pulled out. When this is removed it will be found to measure sometimes a

INFLAMMATORY DISEASES OF THE OVARIES.

sixteenth of an inch in thickness. The cavity left after this will ooze a little, and the organ will appear much shrunk. Nothing more is needed where the apoplexy is small; but where the accumulation is large and its evacuation leaves flabby flaps, these should be trimmed and sutured (Fig. 32). The suture material may be either fine chromic kangaroo tendon or fine carbolized silk. The needles should penetrate beneath the cavity and a contin-

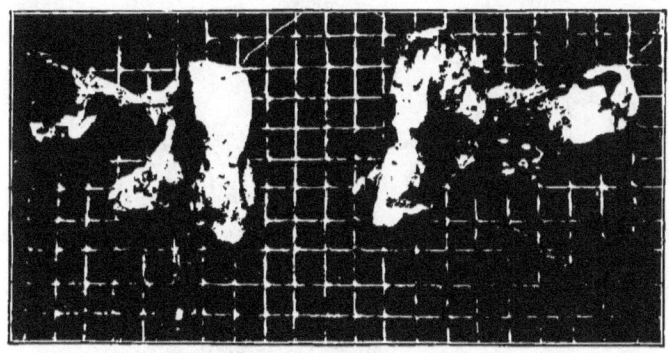

FIG. 33.—Bilateral purulent salpingitis. Bilateral cystic degeneration of ovaries, the right large. Vaginal ablation.

uous suture be used. The ovary is returned, and, after cleansing the pelvis and removing the protecting pads, the opening in the vagina is either sutured or plugged with gauze. Of course, whenever retroversion accompanies either cystic or apoplectic ovary, the cul-de-sac is not to be closed, but is to be treated according to the method described elsewhere (see page 117).

Ovarian Abscess.—The treatment of this is similar to that of pyosalpinx, both as regards palliative operations and extirpation (Fig. 33).

BROAD-LIGAMENT CYST.

Upon examining the broad ligament spread out before a strong light, the various component parts of the par-

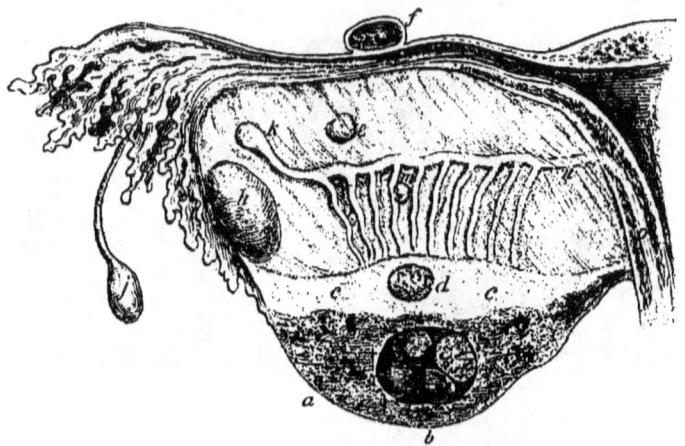

FIG. 34.—Diagram of the structures in and adjacent to the broad ligament. *a*, Framework of the parenchyma of the ovary, seat of *b*, simple or glandular multilocular cyst; *c*, tissue of hilum, with *d*, papillomatous cyst; *e*, broad ligament cyst, independent of parovarium and Fallopian tube; *f*, a similar cyst in broad ligament above the tube, but not connected with it; *h*, a similar cyst developed close to ovarian fimbria of tube; *j*, the hydatid of Morgagni; *k*, cyst developed from horizontal tube of parovarium. Cysts *e*, *f*, *h*, *j*, and *k* are always lined internally with a simple layer of endothelium. *l*, The parovarium; the dotted lines represent the inner portion, always more or less obsolete in the adult; *m*, a small cyst developed from a vertical tube; cysts that have this origin, or that spring from the obsolete portion, have a lining of cubical or ciliated epithelium, and tend to develop papillomatous growths, as do cysts in *c*, tissue of the hilum; *n*, the duct of Gärtner, often persistent in the adult as a fibrous cord; *o*, track of that duct in the uterine wall; unobliterated portions are, according to Coblenz, the origin of papillomatous cysts in the uterus. (After Doran.)

ovarium may be seen either as fibrous cords or as minute tubes (Fig. 34). Any infection passing from the uterus to the iliac glands through the lymph channels in the broad ligament will set up an inflammation in one or more of these embryonic tubes. If one only be inflamed, a single broad ligament cyst will be produced; if more become distended, a multiple cyst is the result. Most women who have suffered infection after abortion and labor will in time develop one or more such cysts of greater or less size. As the cysts grow they spread apart the folds of the broad ligament. They have *no pedicles*. At first, while small and if situated far out in the ligament, they can be moved with the ovary and tube; but when they have grown to touch the side of the uterus, they are always sessile upon the uterus. Their sacs are exceedingly thin and are easily ruptured. The fluid in them is perfectly clear, watery, and of a pale straw color. It is entirely innocent and devoid of harmful properties. Sometimes these tumors are of large size, reaching even to the umbilicus. In growing they displace the uterus laterally. They are never of acute formation, but are of gradual growth.

Symptoms.—Whatever distress attaches to fixity of the uterus and, if the tumor be large, to the presence of a mass, accompanies these growths. There is not the pelvic pain, nor the fever, nor the recurrent inflammation which accompany pus in the pelvis. The history is usually that of a mild degree of infection following abortion or labor. Upon examination there is felt upon one side of the uterus a very fluid tumor, but slightly sensitive. The uterus is firmly fixed to the tumor and may be pushed away from the tumor to one side only. The arch of the base of the broad ligament upon the side of the tumor is destroyed, and the finger when swept away from the cervix on the tumor side appreciates that the tumor and uterus are but one mass. This is an invariable sign of all broad ligament growths, whether fluid or solid, whenever they reach the side of the uterus. By repeated attacks of peritonitis, purulent foci in the ovary and tube may imitate this relation; but such lesions commonly occupy a position further behind the uterus. Broad ligament abscess causes the general symptoms of pus, while ruptured ectopic gestation and broad ligament fibroid are much firmer. The marked fluidity, the thinness of the walls, and the clinical history will usually make the diagnosis clear.

Treatment.—If the growths are small and purely pelvic in location, they are easily treated through the cul-de-sac. But where they are large and extend above the pelvic brim they should be removed through the abdomen. Upon opening the cul-de-sac the diagnosis is easily made. A gauze pad is introduced into the pelvis above the tumor, and the head of the table lowered. The anterior trowel and posterior retractor readily expose the tumor to view. Its surface is smooth and glistening, and through its thin sac the clear fluid is seen. Having inspected the tumor, enough gauze pads are introduced into the pelvis to keep all intestines above the brim, and the patient is brought to a horizontal position. A pair of closed blunt scissors are shoved into the tumor, and its contents escape through the vagina. It is the posterior layer of the broad ligament which is punctured by the scissors. As the scissors are withdrawn the blades are opened so as to make a wide rent in the sac. The pelvis is wiped dry and the finger seeks the opening in the sac. So flimsy are its walls that it is with difficulty found, but when entered its cavity is explored for secondary cysts. These are ruptured. Removing the gauze pads from the pelvis, the surgeon packs the cul-de-sac opening with iodoform gauze which passes just within the cut edges of the vagina. The uterus is replaced and the vagina packed with gauze. I do not *pack* the cyst cavity. It closes spontaneously without artificial drainage, there being no pus present.

The first dressing is made in seven to ten days and repeated as often as soiled. After the second dressing the patient is allowed out of bed.

ADHERENT RETROPOSITIONS.

While this book does not treat of all forms of displacement, there is one so commonly associated with inflammation of the adnexa that I may describe my method of dealing with it through the vagina. I shall exclude from this discussion all cases of congenital displacement and

FIG. 35.—Retroversion with old firm adhesions (Winckel).

shall deal only with those which have been accompanied or caused by either gonorrheic or septic infection, for I believe the congenital cases are incurable (Fig. 35).

The difference between the free and the adherent retropositions is that the latter are complicated by false bands which bind the displaced uterus to the lower and posterior portions of the pelvis, and also commonly present some degree of tubal disease.

Before any attempt at replacement of an adherent

uterus can be made, the false bands of union must be severed. This can be done in one of two ways: either through the abdomen or through the vagina. If the operation is performed through the belly, the fundus uteri is suspended from the anterior abdominal wall (Kelly's method), or else the anterior surface of the uterus is stitched to the upper wall of the bladder (Pryor, N. Y. Jour. Gynec. and Obstet., July, 1893). Few objections can be made to either operation, except that both necessitate an invasion of the abdominal cavity and conservative treatment of the inflamed adnexa is limited in scope. Of the vaginal methods there are two, one of anterior colpotomy (Dührssen-Machenrodt, etc.), which is condemned because it interferes so often with subsequent pregnancies; and the other, the operation I have for years been performing. I have been struck with the invariable observance of one of two rules in all operations which succeed in keeping a retroposed uterus forward: either this is accomplished by fastening the corpus uteri forward, or else by fixing the cervix high and backward so that the intra-abdominal pressure will force the body of the uterus forward. This latter is the way a pessary acts, and this is the idea embodied in my operation.

Operation.—The patient is prepared locally and generally as for a capital operation. I begin the operation with a curettage. The cul-de-sac is then opened (see Exploration). Upon entering the pelvic cavity I make a careful digital exploration. If I find a pus focus. *I abandon all further attempts at replacement by the vagina*, and treat the case as one of suppuration. But if I find any condition of the adnexa that will not require their removal (see Conservatism), I continue the operation. Occluded tubes are opened and other adhesions are severed. The pelvis is then wiped dry, and a gauze pad inserted. The patient is tilted into the Trendelenburg posture and the gauze pad is removed. The uterus is packed with iodoform gauze. The operator selects a piece of iodoform gauze wide enough to fill the vaginal opening and about one and a half inches long. This is inserted *just within* the edges of the vaginal rent. Over this enough strips are placed to fill the incision in the vagina. This gauze plug, together with the uterus, is next replaced. It is easily done, as the patient is head down and the intestines have left the pelvis. Holding the uterus in position, by means of the trowel or any depressor pushing against the cervix, pieces of gauze are inserted to the sides of the cervix and in front of it until the vagina is filled to the margin of the levator ani muscle. The operator now takes a stout roll of gauze as thick as his thumb and as long as the width of the distended vagina, usually two inches. This I call my gauze pessary. One end of this is introduced in front of one side of the cervix, just behind the levator ani fibers, and the other end is pushed into a similar position on the other side. This plug will lie transversely across the vagina and in front of the cervix. (Fig. 36). It will prevent descent of the cervix even in face of the most violent vomiting. The uterine packing should be so arranged that it can be removed without disturbing this anchoring plug.

A self-retaining catheter is introduced and is emptied every two hours for two days. The bladder is then irrigated with boric acid solution and the catheter withdrawn. The uterine packing is now removed without disturbing the vaginal. In seven to ten days the patient is placed in Sims' position. All dressings are removed and replaced exactly as were the first. The operation will fail unless the supporting plug is properly inserted. This is as important as the suture in other operations. The second dressing is applied a week later, is painless, and after it the patient sits up. I keep up these dressings as long as there is any raw surface at the vaginal vault; the supporting tamponade I use for six weeks. The woman is then allowed intercourse.

If at any time the dressings are so applied that they allow of descent of the uterus, they have been improperly inserted. The cervix must be kept high and backward until the cul-de-sac opening closes and the post-cervical scar has contracted. The operation leaves the corpus uteri perfectly free. Pregnancy resulting after the operation is uninterrupted, and labor is normal. Lacerations and disease in the cervix and perineum are to be cor-

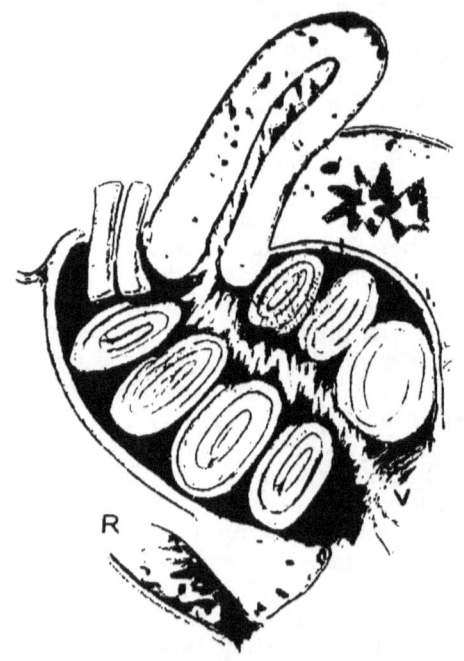

FIG. 36.—Showing schematically the position of the dressings in the cul-de-sac operation of replacement.

rected after the patient has recovered from the replacement operation, and are made purely to supplement the first operation. The rules governing these plastic operations are the same as apply after hysterorrhaphy, etc.

The operation in my hands takes the place of all other

operations. It has a wider range of application than any other procedure, and can be used in all cases not presenting pus. When the retroposition is accompanied by occluded tubes, by hydrosalpinx, by cystic ovaries, etc., this is the preferable operation. But when pus is present in either ovary or tube removal of this and replacement can only be accomplished by laparotomy.

BROAD-LIGAMENT ABSCESS.

This rare condition almost invariably follows labor or abortion. The infection passes along the lymph streams between the folds of the broad ligament, and causes suppuration there. The pus forms very slowly usually. In the epidemic of puerperal fever which occurred in New York in 1881–82, the author saw a great many of these cases, but they are now comparatively rare. I have met with but six in the last 1,000 clinic cases. As the pus accumulates, it separates the folds of the broad ligament. The bladder in front prevents much bulging anteriorly, so the greater part of the distention of the broad ligament is posteriorly. As this grows larger, the peritoneum is stripped from the posterior surface of the uterus and is lifted up; the peritoneum of the pelvic floor behind the broad ligament is also lifted, and the masses may be so large as to reach Poupart's ligament. The fluid is essentially extraperitoneal. It is suppuration in continuity of tissue, and is far different in all its bearings from suppuration in a preformed sac (pyosalpinx). Coexistent with this formation, there is a great amount of peritonitic effusion about the broad ligament. There may also be a pyosalpinx or ovarian abscess present. After the abscess reaches a large size, the gross lesion presented is of an abscess cavity lying upon the pelvic floor, to one side of which is the displaced uterus, and above which lies the matted mass of omentum and intestines. In rare cases the abscess is bilateral. In such the pus may extend in front of the uterus and between the bladder, so that the two abscesses communicate.

Symptoms.—These are at first not suggestive of broad ligament abscess. After a long attack of continuous pelvic inflammation, in the course of which there have been many rigors and violent fluctuations in temperature, this condition may be suspected. Upon examination, the uterus is found crowded up high and to one side. It is sometimes so displaced that the cervix cannot be felt. Extending from the side of the uterus to the lateral pelvic wall is a large mass, tense and fluctuating. This mass is sessile upon the uterus, $i.\ e.$, there is no sulcus between the mass and the uterus. It is immovable and fixes the uterus. It projects in all directions when large, and can be felt behind the bladder, above Poupart's ligament and deep in the pelvic floor. There are evidences of a severe type of pelvic inflammation. The bladder is capable of holding but a few ounces of urine when the abscess is large, and the lumen of the rectum is almost closed. Upon rectal examination, the mass is found apparently attached to the rectum if it has stripped up the pelvic floor. About the only conditions simulating this are, dermoid cysts, ectopic gestation ruptured and septic, and broad ligament cyst. But the history of a labor or abortion, long-continued sepsis, and a gradually enlarging tumor, which is always sessile, even when small, and which undoubtedly occupies the broad ligament, will render the diagnosis clear. When the accumulation is small, the finger readily enters the incised cul-de-sac. The enlargement is found to be upon one side of the uterus, and the posterior wall of the broad ligament bulges backwards. A slight pressure against the mass suffices to evacuate the pus, rendering the diagnosis clear. With this pus formation, there has been much peritonitis, and the examining finger evacuates the lymph and serum produced by this. The pelvic viscera, where they can be reached, are found matted together. The ovary and tube upon the affected side are raised high in the pelvis. When the abscess is large, the finger cannot be made to enter the cul-de-sac at all; and upon incising the vaginal mucosa, the finger enters a cavity of loose cellular tissue which bleeds freely. This is produced by the abscess lifting the peritoneum from the pelvic floor. After inserting the finger behind the uterus up to the level of the internal os, it will enter the pus sac at once, or will find it if turned laterally toward the fluid mass.

This lifting of the pelvic peritoneum is characteristic of all large broad ligament accumulations.

Treatment.—All these accumulations should be treated through the vagina. They should be opened through the posterior cul-de-sac and evacuated. For this purpose the fingers alone are to be used after the vaginal wall is incised, as the position of the vessels is not constant. If the cul-de-sac is entered before the abscess is emptied, it, as well as the abscess cavity, must be packed with gauze. If the examining finger enters at once into the pus cavity, it is to be widely stretched and packed. The after dressings are governed by the amount of discharge and the temperature. It is wise to curette the uterus before opening the cul-de-sac. After dressing the abscess cavity the uterus is to be packed with iodoform gauze, which in two days is withdrawn.

DIFFUSE PELVIC SUPPURATION.

This must not be confounded with primary purulent peritonitis. There has been suppuration in either the ovary, the tube, or the broad ligament. Accompanying this there has been a virulent form of peritonitis, and a great outpouring of plastic lymph has ensued. Sometimes this lymph breaks down into pus; in other cases the original pus focus leaks into the lymph masses. As a result the pus has ceased to be confined in either tube, ovary, or broad ligament, but has wormed its way between adherent lymph planes, omentum, and intestines.

More lymph is produced, and wider burrowing of pus ensues, thus presenting a picture of indistinguishable organs within and between which are pus pockets and connecting sinuses. This is diffuse suppuration.

Symptoms.—The history is usually one of prolonged suffering, recurrent attacks of peritonitis, emaciation, and hopelessness. The woman is practically bed-ridden. Upon examination the uterus is found firmly imbedded in a mass of exudate. The uterus, ovaries, tubes, and other pelvic organs form one dense conglomerate mass. The diagnosis from small fibroid or ruptured ectopic pregnancy with pus is impossible. Broad ligament cysts, ovarian tumors, simple fibroids, etc., do not present the immobile sensitive uterus, profound sepsis, emaciation, and mal-nutrition which accompany diffuse suppuration. Nephritis and phthisis are common accompaniments. I have usually found the rectum permanently distended in these cases. It cannot contract.

Very often sinuses form between the bowel and the pus foci, affording a temporary relief when the pus escapes into the gut, followed by great increase in the lesions from contamination by bowel filth.

Upon opening the cul-de-sac, the finger at once evacuates pus lying free in the pelvis. The exploration is purely digital, and as the finger maps out the various organs it enters pocket after pocket of pus. The livid lymph-covered intestines are found low down usually, pressed down by tympanites, and tend to protrude into the vagina when freed. Above the uterus and adnexa is an impenetrable dome of matted intestines and omentum.

Treatment.—Many of these women are so critically ill that a radical operation is contraindicated. The first step is usually to inject a quart of sterile and filtered normal salt-solution into the elbow vein. The uterus is curetted. The cul-de-sac is opened, and all pus pockets emptied. The pelvis is wiped dry with gauze. Irrigation should never be employed for this purpose, lest the pus be washed into the higher pelvis and abdomen. After evacuating all the pus cavities and thoroughly cleansing the pelvis, the dressing is made. The long perineal retractor draws down the posterior vaginal wall while the trowel lifts up the uterus. An abundance of light is by this means thrown into the pelvis, and the gauze can be inserted between two smooth metal planes. Each piece of iodoform gauze is three inches wide and a yard long. It is folded many times, and is inserted up to the level of the fundus uteri. With a lateral retractor this piece is pulled to one side while another is inserted, and progressively the pelvis is filled.

The first dressing will not require narcosis, and should be made as soon as the temperature rises. The opening is kept carefully packed until it is closed by granulation. Every pocket of pus is sought out and entered. Its sac is widely opened by the finger, and after swabbing it dry it is packed. The operator must not fail to insert a stout gauze drain into every pus pocket.

This operation is purely palliative, and a relapse is to be expected. But before this occurs the case can be carefully prepared for a radical operation. After the general condition has been improved the uterus, ovaries, and the tubes are to be removed through the vagina by hemisection.

ANESTHETIC.

The vaginal operations do not demand that complete physical relaxation which is essential in laparotomy. An incomplete narcosis is sufficient, and chloroform again becomes the preferable anesthetic. I have extensively tried the Schleich mixture, but have returned to chloroform and ether. I always administer chloroform upon an Esmarch mask, and usually precede it by a hypodermic of strychnin, gr. $\frac{1}{50}$. Ether I give through a Sims' inhaler, sterile gauze being employed to hold the anesthetic. The cones with bag attachments I have thought conduced to the occurrence of pneumonia.

CURETTAGE.

The patient, having previously been prepared, is placed upon the back in the lithotomy position. The perineum is retracted with a short speculum, or the cervix may be exposed by means of a bivalve speculum. If the latter be used its blades must be short. The cervix is next seized with a pair of blunt bullet forceps fastened into the anterior lip, and the cervix is gently drawn down. This at once brings the cervix nearer and straightens the canal very materially. The direction and depth of the canal are determined by the sound. The dilator is introduced, the cervix being still firmly held (Fig. 37). In selecting a dilator the operator should avoid those operated by screws. These are positively dangerous, inasmuch as the force can not be released if tearing begins, so that rupture of the cervix may be produced. Some modification of Sims' instrument is best. If the cervix is so stenosed that the dilator cannot be introduced, I do not hesitate to enlarge the canal with a blunt bistory, cutting bilaterally. The dilatation is done progressively, the force being intermittent and the dilator turned a little from side to side. The dilatation in all cases reaches a half inch, and greater space is secured in many cases. The larger the cavity to be scraped, the more open should be the cervix. If the dilatation is properly done and a good deal of traumatism inflicted upon the cervix, the cervical ganglia are obtunded, and uterine contractions with expulsion of the dressings do not follow the operation. Having dilated the cervix the uterus is to be curetted. I use the sharp instrument of Sims, and prefer a small one for hard uteri and a large one for soft organs. The curette is introduced to the fundus gently, and the force is used in withdrawing the instrument. The pressure is made all along the instrument, and the cervix must not be used as a fulcrum. The operator proceeds all around the inside of the uterus,

FIG. 37.—Demonstrating the method of dilating the cervix preliminary to curetting the uterus.

paying particular attention to the lateral angles and tubal openings. The fundus is scraped by sweeping the curette from one tubal orifice to the other several times. The uterus may now be irrigated with saturated solution of boric acid or normal salt-solution. The ordinary fountain syringe is the best irrigator (see page devoted to sterilization). I employ the Fritsch-Bozeman double current catheters. If the débris removed by the curette are small they are readily washed out, but large plugs of tissue must be wiped away. To accomplish this the uterus is packed with gauze and the dressing is made to revolve within the uterus by means of a tampon screw. When it is withdrawn, all portions of membrane are caught in the folds of the gauze. I have abandoned irrigation of the uterus where the organ is small and a small irrigating tube must be used. Instead I swab out the cavity by iodoform gauze. The sole object is to remove all débris produced by the curette. In many irrigated cases where I have opened the cul-de-sac, I have found fluid in the pelvis which was blood-stained. I have thought that perhaps in certain cases of small uteri the irrigating fluid may have escaped into the pelvic cavity through the tubes. Having cleansed the uterus, it is next packed with gauze. I use for this purpose a stout metal applicator which is slightly curved. The gauze is folded over the end, and is fed into the uterus by means of successive holds with the applicator. The tampon screw can be used for the same purpose (Fig. 38). The organ should be completely filled. The object of this is to have within the uterus a sufficient amount of dressing to exert pressure, to absorb all discharges, and to act as a protection to the repair-cells while they are forming a new membrane. The first few days are the most critical in this matter. If the repair is started properly, it will proceed to the formation of a histologically perfect membrane. But if the first emigrated cells are destroyed, either by iodin, carbolic acid or by infection, a distorted endometrium results, which will produce painful menstruation and sterility. Asepsis of the most precise nature is the only method to bring success.

It will not suffice merely to remove the endometrium; the zinc pencil will do that. But the removal and treatment must be so done as to insure a reproduction of a perfect new membrane. That is impossible when even mild sepsis follows the operation. As antiseptics destroy cells they should never be used within the uterine cavity, for dead cells disturb healing and furnish the most propitious culture medium for germs. Antiseptics have no place in cavity work like this. Some of the worst cases of dysmenorrhea and pelvic neuritis I have found in women who have been curetted by careful men, men perfectly cleanly in their methods, but who, believing the endometrium to be a mucous membrane, have painted it with carbolic acid, thereby promoting the production of scar tissue within the uterus without a trace of lymphoid elements. The operation of curettage must be done with a knowledge of the anatomy of the uterus and a consciousness of its function.

The vagina is usually packed with iodoform gauze. If uterine cramps follow the operation a suppository of ext. opii and ext. belladonna, each gr. $\frac{1}{2}$, may be given. The vaginal gauze is removed on the third day and the uterine packing withdrawn. The uterus is neither irrigated nor again packed, but the vagina is packed again, and the woman allowed out of bed, provided the curettage has been done for an uncomplicated endometritis. The bowels are moved on the third day. The vaginal dressings are removed once in four days for two weeks, and then all treatment ceases. A new endometrium forms in from four to six weeks. Douching and amatory approaches from the male are forbidden until after the period following the operation. Such is the usual course and the usual operation.

Time for Operating.—Preferably one week after menstruation is the elective time. The special conditions under which the operation is done will modify this and may be referred to.

Infected Cervix.—When I operate in acutely inflamed cases and when the cervix is the seat of gonorrhea, I

FIG. 38.—Packing the uterus with iodoform gauze.

always paint its cavity with pure carbolic acid, both before and after dilating. Equal parts of tinct. iodin and carbolic acid make a powerful antiseptic. In using these caustics care must be exercised not to allow of the passage of a particle into the uterine cavity above the internal os. The carbolic does not cause sloughing in the cervix, because its lining membrane is a very dense mucous structure. I consider it exceedingly important to sterilize the cervix in all gonorrheal cases. The best means of doing this is by means of carbolic applied before the glands are emptied by pressure and after.

Repeated Irrigations.—Whenever I have to deal with a *large* acutely infected uterus, and especially if there be peritoneal or adnexal lesions beginning, I deem it necessary to irrigate the uterus when I withdraw the uterine packing. In these cases the infection is usually deeper than the surface, and repeated washings with boric acid may be necessary during the process of repair. After washing out the uterus a filament of gauze is introduced to the fundus and the vagina again packed. The filament of gauze ensures an open cervix. I remove this second dressing in three days, and am governed by the appearance of the discharges as to whether I shall repeat the washing or not. If pus, or even broken-down material in quantity, follows the withdrawal of the second gauze, I again wash and pack as a precaution against re-infection. Cases which are infected post abortum or post partum always need at least one renewal of the washing and uterine drain.

Gonorrheal cases and those infected by uterine tinkering sometimes require this washing.but no packing after the first dressings are removed. It is almost needless to caution the operator regarding cleanliness all during the treatment. The sterilization must be as complete at dressings as at the operation.

The two causes for infection after curettage I have found to be in the faulty removal of débris and the application of escharotic antiseptics. So long as the operation performed has these two attributes, of course the question will arise, Does gauze drain or not? Thus far I have not found it necessary to concern myself with this, but I may answer the question. Gauze *does* drain, for the vaginal packing is often wet through with secretions such as are found in the uterine packing.

QUANTITY OF GAUZE IN UTERUS.—The girl's uterus measuring three inches will hold a strip of gauze one inch wide and a yard long. The uterus aborted at the third month will contain a strip four inches wide and a yard long. The full term uterus will receive a roll of gauze one yard wide and five yards long.

THE INSTRUMENTS.—I prefer the short specula of Jackson (Fig. 39). They are simple, depress the perineum properly, and are as useful when laparotomy or vaginal hysterectomy is done as in curettage. The traction forceps should be very dull (Fig. 40), so as not to tear the tissues. I can see no necessity for the multitude of dilators offered for sale (Fig. 41). The instrument of Sims', roughened as I have had done, is sufficient. Dilatation

FIG. 39.—Jackson speculum.

FIG. 40.—The author's blunt bullet forceps.

by graduated bougies is always imperfect. The bougies are shoved in against the force of the traction forceps, and

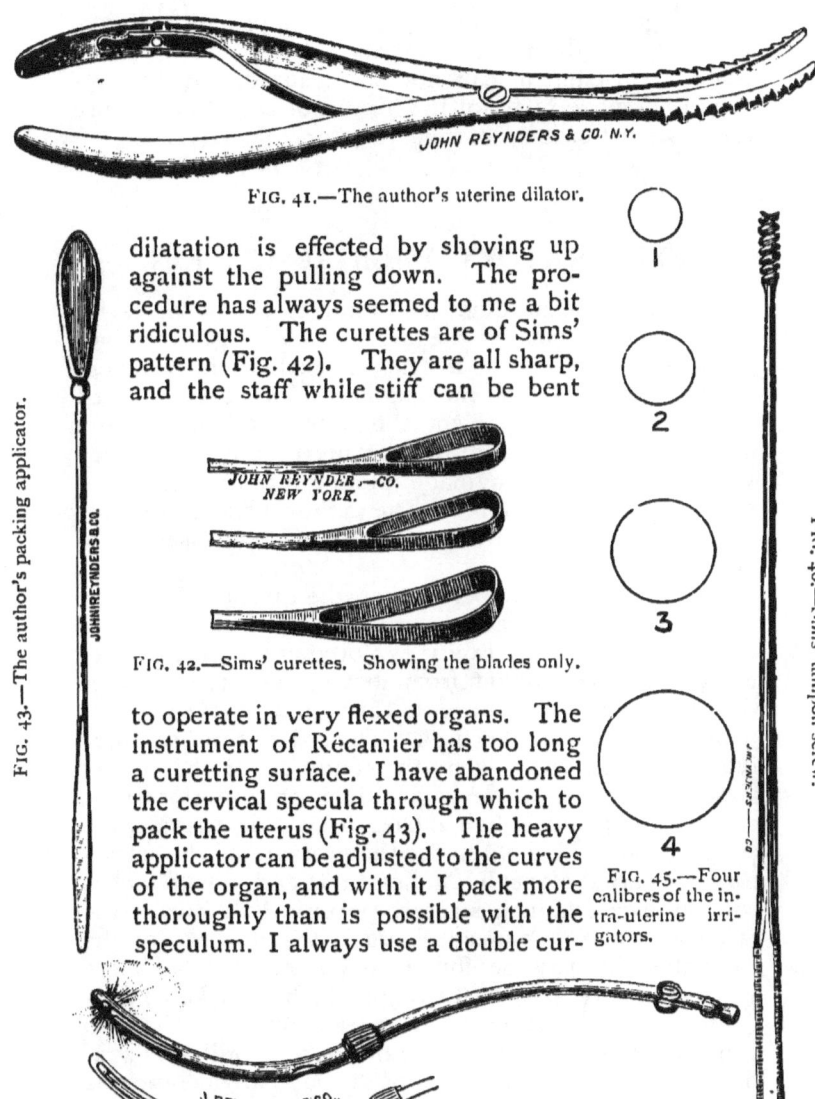

FIG. 41.—The author's uterine dilator.

dilatation is effected by shoving up against the pulling down. The procedure has always seemed to me a bit ridiculous. The curettes are of Sims' pattern (Fig. 42). They are all sharp, and the staff while stiff can be bent to operate in very flexed organs. The instrument of Récamier has too long a curetting surface. I have abandoned the cervical specula through which to pack the uterus (Fig. 43). The heavy applicator can be adjusted to the curves of the organ, and with it I pack more thoroughly than is possible with the speculum. I always use a double cur-

FIG. 42.—Sims' curettes. Showing the blades only.

FIG. 43.—The author's packing applicator.

FIG. 45.—Four calibres of the intra-uterine irrigators.

FIG. 46.—Sims' tampon screw.

FIG. 44.—Fritsch-Bozeman double-current irrigating tubes.

rent irrigating tube (Fig. 44). It is improper to inject irrigating fluid into the uterus with a bulb syringe. Such an instrument can not be cleansed, and no provision is made in it for the return of the fluid. I use the Fritsch-Bozeman uterine irrigator.

EXPLORATORY VAGINAL SECTION.

The bar to a thorough inspection of the pelvic cavity through the vagina is the uterus; and a great embarrassment experienced in the procedure is prolapse of the intestines into the vagina. If a posture can be secured which will prevent the latter, and an incision adopted which will remove the uterus out of the way without injuring it, vaginal exploration of the pelvis will supersede the abdominal. The author believes that his procedure secures both the desirable factors essential to success.

It must be remembered that the distance from the vulva to the cul-de-sac is even less than from the abdomen. Therefore the cavity explored from below is not as deep as when sought from above. The ability to see the pelvic structures through the vagina is then dependent upon the space secured. The space is not so much limited by the vulva as by the condition of the tissues about the cervix. If the vaginal incision posterior to the cervix is one and a half inches from side to side, the elastic tissue will yield under the pressure of the retractors to make the opening at least one and a half inches wide by over two inches antero-posteriorly. But in the rare cases of pronounced sclerosis the elasticity of the vaginal vault may be found so limited that sufficient space cannot be secured through which to make an adequate visual inspection. The operator will then have to depend wholly upon his sense of touch. Still this contingency is not as often met in the vaginal operation as in the abdominal.

EXPLORATORY VAGINAL SECTION.

Operation.—The local and general preparation of the patient will be found on pages 161 to 163. The patient is placed upon the (Fig. 47) table in the lithotomy pos-

FIG. 47.—The cul-de-sac is opened. The posterior vaginal wall is held down by the retractor, while with the trowel the uterus is shoved up against the bladder. The space obtained is estimated by comparing the length of the operator's index finger with the distance between the blades of the retractors. In this case it was 2¼ inches.

ture, with the ischial tuberosities over the edge of the table. The perineum is retracted by a short Jackson speculum, and the uterus is pulled down. The uterus is curetted and swabbed out, but not packed with gauze. The vagina is wiped dry. Upon shoving the cervix upward a fold will be seen to form just opposite the cervico-vaginal junction (Fig. 48). The vagina is incised here, scissors being used for the purpose. The scissors

138   PELVIC INFLAMMATION.

cut through vaginal mucous membrane only. The incision is commonly an inch long and extends to the lateral borders of the cervix (Figs. 49, 50). There now remains

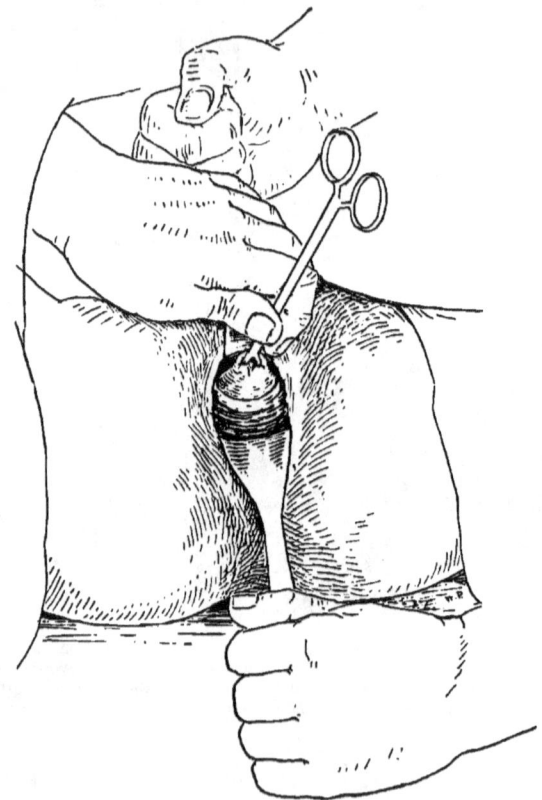

FIG. 48.—The fold behind the cervix which lies over the cervico-vaginal junction is well shown. The vagina is to be incised here (from life).

but one layer of tissue to sever,—the peritoneum. The uterus is held firmly down, and the operator pushes his index finger into the cul-de-sac. In doing this he is careful to keep the point of the finger accurately

EXPLORATORY VAGINAL SECTION.

in the middle line and pressed up against the posterior uterine wall. If after pushing the tissues up to the level of the internal os the finger has not entered the peritoneal cavity, the point of the finger is directed backwards

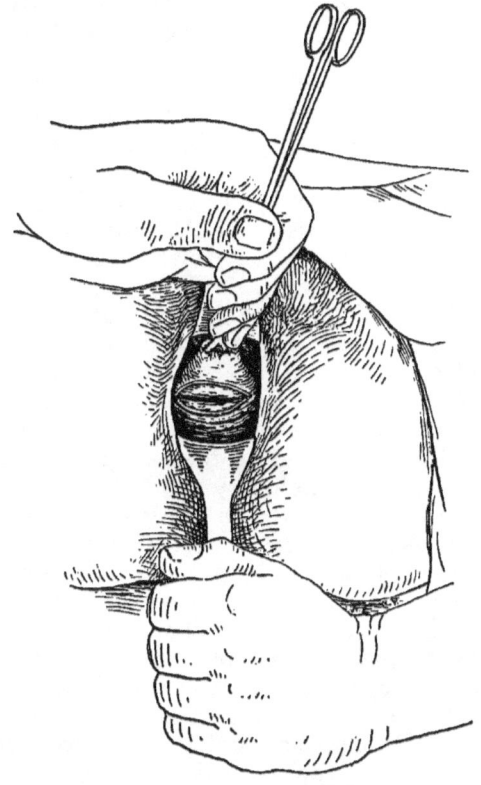

FIG. 49.—The vagina is incised, and the point at which the peritoneum is reflected from the uterus is shown as the deepest part of the cut. The peritoneum is to be torn through at this point (from life).

and pushed into the cavity. If the peritoneum is very thick it is caught with toothed forceps and incised with scissors. Commonly serum escapes when the cavity is entered.

In making the incision one small vessel is severed,—the azygos artery of the vagina.

It requires forcipressure very rarely, being an insignificant vessel. Having entered the pelvic cavity a gauze pad, to which a string is attached, is introduced. While the operator washes his hands, an assistant lowers the table into the Trendelenburg position. At once all unattached viscera leave the pelvis.

FIG. 50.—1-2, The anterior incision, used when hysterectomy is to be performed. 3-4, The posterior incision, employed when the pelvic contents are to be examined and the viscera treated conservatively.

The operator now inserts his two index fingers into the rent, and upon separating his hands the incision is spread laterally (Fig. 52). This tear takes place in the line of the incision. A careful digital examination is now made of the pelvic contents. The finger glides up along the smooth posterior uterine wall as high as the fundus and is then swept laterally over one cornu and tube. The ovary and tube upon one side are carefully palpated. If tender adhesions are met with, they are torn with the finger. Unless pus is suspected, the effort is made to free the ovary and tube from adventitious union. The operator remembers that his finger has entered *below* the plane of the bases of the broad ligaments, and that his manipulations are behind the broad ligaments, upon their posterior surfaces. At once this will indicate to him the method of separating adherent adnexa. In doing this the finger is moved between the surfaces of union from the side of the uterus upward and outward, a sort of lifting motion being made. All the time the adnexa are being manipulated, the uterus is firmly held down with the bullet forceps. The pelvis is now wiped free from blood. If firmly adherent adnexa, or cystic accumulations are met with, it is better not to complete their separation before inspecting them. Inspection of the pelvis is next made. A medium Péan

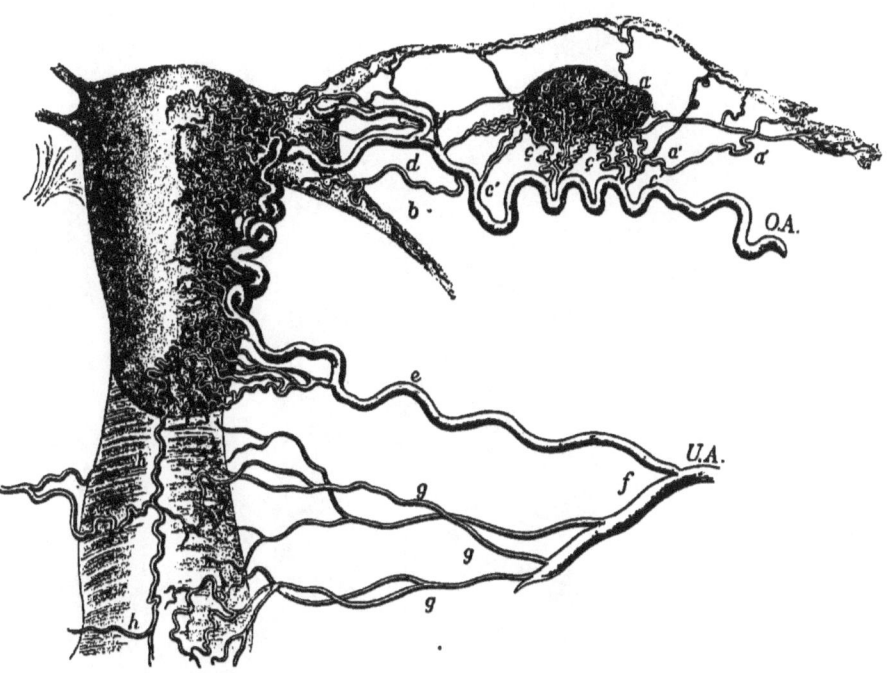

FIG. 51.—Arterial blood supply of the uterus and adnexa: *O. A.*, ovarian artery; *a′, a′, a′*, branches to ampulla of Fallopian tube; *c′, c′, c′*, branches to ovary; *c*, branch to fundus; *d*, branch anastomosing with uterine; *b*, branch to round ligament; *e*, uterine artery; *g, g, g*, vaginal arteries; *h, h*, azygos artery of vagina.

EXPLORATORY VAGINAL SECTION.

retractor is introduced (Fig. 53), and the perineum, vagina, and posterior edge of the incision are held down by it. The cervix is loosed from the grasp of the bullet forceps, and a Péan-Pryor trowel is inserted behind the uterus. The soiled pad is now removed, and several clean ones are inserted. The uterus is pushed up behind the symphysis and out of the pelvic cavity by the trowel. This is the very essence of the procedure, for by it the

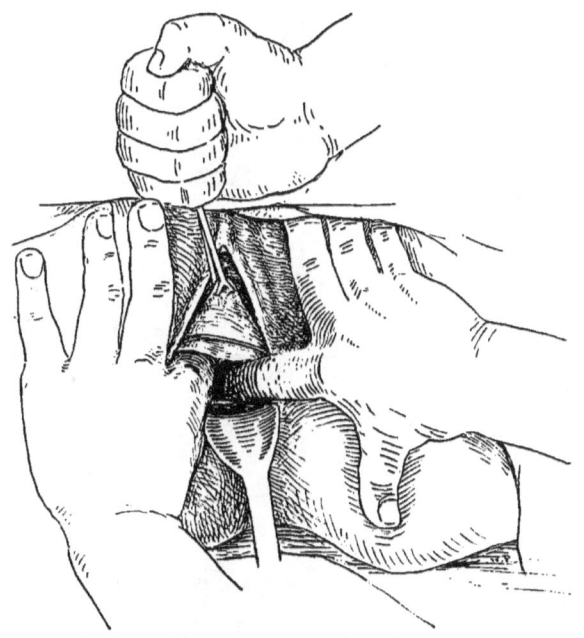

FIG. 52.—The index fingers are inserted into the opening in the cul-de-sac, and the incision is enlarged by blunt tearing with the fingers (from life).

obstructing uterus is lifted out of the way. By dexterous manipulation of the trowel the adnexa of first one side and then the other are exposed to view. When seen they may be grasped with Luer's forceps and brought down into the vagina, where they may as readily be operated upon as is the cervix in plastic work (Fig. 54).

PELVIC INFLAMMATION.

Pelvic exostoses, adherent vermiform appendix, rectal cancer, ectopic gestation, both unruptured and ruptured, ovarian cystoma, ovarian sarcoma, uterine fibroids, hydrosalpinx, pyosalpinx, cystic and apoplectic ovaries, occluded tubes, dilated ureters, and in fact every form of pelvic disease I have seen are most of them treated through such an incision. By relaxing somewhat the perineal

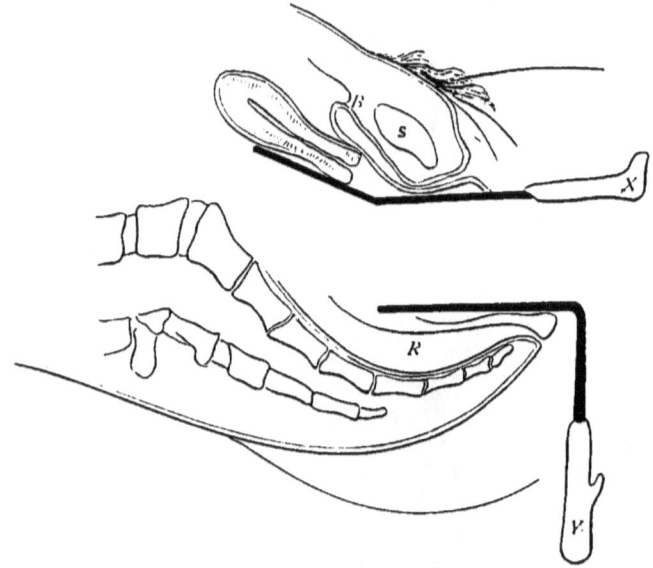

FIG. 53.—The uterus is held up behind the symphysis (S) with the bladder (B) by the trowel (X), while the rectum (R) and the posterior vaginal wall are pulled down by the retractor (Y).

retraction and forcibly pushing up the uterus and bladder the ureters are made tense and appear as curved ridges beneath the lateral pelvic peritoneum. The space gained between the trowel and retractor is nearly that made by separating the fingers. The further operative treatment depends upon what the inspection reveals. Suffice it to say that I pack the uterus with gauze, remove the gauze pads, insert gauze into the pelvic opening or close the opening with fine silk, and replace the uterus. The cul-de-sac is entered in two minutes; the entire procedure occupies but ten. I commonly employ a partial chloroform narcosis, as complete relaxation is not necessary.

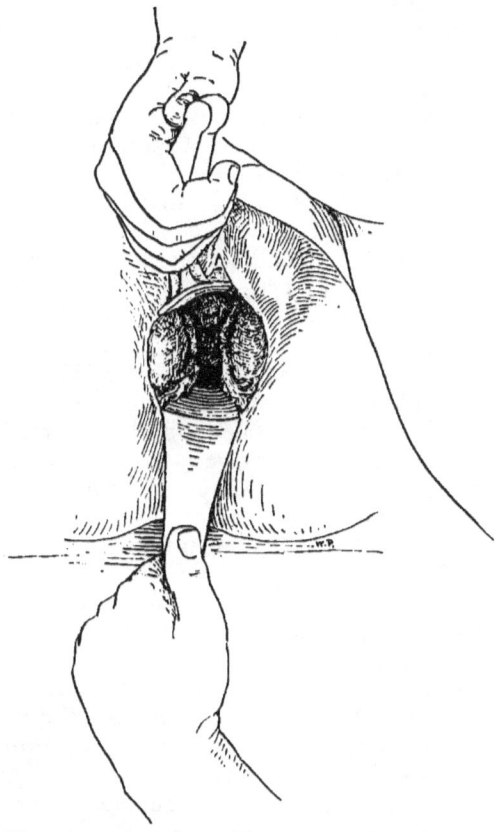

FIG. 54.—The adnexa have been freed, and are brought down into the vagina. Above them are coils of intestine (from life).

There are but two conditions in which this method of exploration is not a completely successful one, and in them the indication for radical operation is so clear that exploration is unnecessary. I refer to ectopic gestation ruptured into the broad ligament, and to intraligamentous fibroid tumors. In all forms of adnexal inflammatory disease and ovarian neoplasms I have found it eminently satisfactory.

All cavity operations are in their first stage exploratory, and my operation occupies that position with regard to future work.

I will contrast the abdominal method and that by section in front of the uterus with my operation. In abdominal section the following anatomical layers are severed: skin (usually infected), fat, fascia, muscle, and peritoneum. In anterior colpotomy (separating the bladder from the uterus) the vagina is severed, the tissues uniting the bladder and cervix (pericervical) are cut, the peritoneum is cut. In my operation two layers are severed, the vaginal wall and the peritoneum.

In abdominal section many small vessels are cut often requiring ligation. In anterior colpotomy the large branches of the superior vesical and uterine arterial anastomosis are severed and require ligature. In my operation no vessel needs more than a few minutes forcipressure. After laparotomy a number of sutures are required to close the wound, and few operators are agreed how this should be done.

With anterior colpotomy the bladder must be again sutured to the uterus by a complicated method. In my operation no sutures are needed.

There is little danger of wounding any important structure during abdominal section, but there is great danger of wounding the bladder in doing anterior colpotomy. There is no possible risk run in my operation of wounding any organ, as the finger does the operating after the vaginal mucosa is severed.

Mural abscess, hernia, inter-intestinal adhesions, and adhesions between the scar and viscera, result often from abdominal section, and an ugly scar is left as a reminder of an unpleasant experience. After anterior colpotomy the uterus is held low in the pelvis and can not readily rise because of thickening in the pericervical tissues. Hence pregnancy is often interrupted. No sequelæ follow my operation, and the uterus is not limited in upward movement.

The entire pelvic contents can be seen by abdominal incision; anterior colpotomy necessitates pulling the uterus down into the field of vision, and hence nothing is seen until dragged into the vagina; the whole pelvis can be explored by means of my procedure. Abdominal section necessitates a profound narcosis, a partial one suffices for the operation I advocate.

To sum up, abdominal section and anterior colpotomy inflict needless traumatism, furnish no drainage space, and are most complicated in every way, while my operation merely requires a special table and instruments to be simply and easily done, and with ample drainage space for all discharges.

In the discussion of "conservatism" diseased organs will be mentioned as cured without removal which would have been sacrificed had laparotomy or anterior colpotomy been done.

The after treatment is that of the "operation for adherent retropositions," which see.

If this method of inspecting the pelvis accomplished nothing else, it should have an accepted place in our procedures as a means of clearing the diagnosis of a suspected ectopic gestation. It is no longer necessary to wait for symptoms of hemorrhage and exsanguination. In three minutes or less the diagnosis can be made, and with no risk. A special table is not necessary for ordinary vaginal work. One may be improvised which will give an angle of 60° by sawing off two legs of a stout kitchen table. The shoulder braces can be arranged by boring two rows of parallel holes down the centre of the table into which pins may rest. Against these the shoulders may be supported, or assistants may support the body while in position. This I have often done.

CONSERVATIVE TREATMENT.

General Considerations.—When we attack diseased adnexa either through the abdomen or through the vagina, up to a certain stage the operation is in all cases exploratory: for, be the presumptive evidence what it may as corroborative of the diagnosis, the absolute diagnosis can be made only after inspection of the diseased structures. When operating through the belly, meeting with a hydrosalpinx or a large cystic or apoplectic ovary, or a small broad ligament cyst, removal seems rather too severe for the pathological lesions. Still it would appear unsafe to return the organs after evacuating fluids and inflicting extensive traumatism.

So long as pus foci only are removed through the abdomen, without drainage, little criticism can be passed. Still, even pus tubes can be conservatively treated by another method. But when lesions of the adnexa are treated by extirpation, which lesions by no possibility can endanger life, not only the operator, but his art as well, is brought into disrepute. There is legitimate ground for debate regarding the propriety of treating pus foci by any method other than extirpation through the belly. But there is no excuse for sacrificing organs trivially affected, when there is at our command a method difficult to be sure, but perfectly safe, which radically cures and yet saves.

But local conservatism must not be carried too far. There is a broader conservatism which seeks the preservation of the general health at the sacrifice of even important organs. To illustrate, the attempt to benefit permanently the condition of a woman suffering from diffuse pelvic suppuration by any operation other than the most radical is absurd. Equally so is it to expect conservative surgery of any sort to relieve the pains of genital sclerosis affecting uterus, ovaries, and tubes. I cite these two extremes to demonstrate the necessity for a careful differentiation before determining upon a particular operation.

I suppose in no field of surgery is our art more hampered, modified, and with reason influenced by the extraneous circumstances surrounding the patient. A young girl of nineteen with large pus tubes is of course better without them. But to check the new function of menstruation at its inception is to create a very unhappy woman. Therefore, here the vaginal conservative operation is pre-eminently indicated; whereas, in a woman of forty, with children, the question would not be considered except as bearing upon risk to life. The young woman again may have pus tubes due to an abortion, and be phthisical. No matter what his sentiments and desires, the surgeon will be governed by his knowledge of the effects upon such a constitution of prolonged suppuration, and of the great improvement induced in the general nutritive functions by artificially inducing menopause. Here the radical operation is always indicated.

The ability to determine upon the proper operation, whether radical or conservative, depends upon so many factors entering into the surroundings of the patient and her station in life, the equipments of the operator and the place in which he works, that I can not possibly foresee them all. The statements I have here made are occasionally modified, and I can but give a description of my usual course in dealing with the different conditions.

But what argument can be offered against the point of view of the conservative man? I take it, none. Then criticism can justly lie upon his results only. And of these I will now speak.

Thousands of cystic ovaries are every year removed through the abdomen. The patient may be high-strung and nervous. She has pre-menstrual pain in the ovarian region; her flow is scanty; her uterus is small and anteflexed; she has severe pelvic tenesmus; she is hysterical and is given to introspection and self-pity; she never has fever; the ovaries are low down and sensitive. Commonly, the woman is in advanced maidenhood. An operation is advised and accepted. Upon opening the belly, no adhesions are found, unless intra-uterine tinkering has produced peritonitis. The tubes are normal. The ovaries are large and filled with cysts, some small, some large. The ovarian capsules are thick and tough. What is to be done? To enucleate all the cysts and suture the cut surfaces of the ovaries is eminently proper. But the same train of nervousness, introspection, regrets, remain, only now she has a belly-scar to study twice a day, when she dresses and undresses. This she will watch carefully for hernia. But the woman is not physically cured. To remove the ovaries is to add the distress of the artificial menopause to her other symptoms, and make her still more hopeless. Although she has never expected to have children, she is now relegated to the class of "the spayed." This picture is not one bit overdrawn.

Through the vagina these cystic ovaries can be successfully treated, cysts emptied, and cut surfaces sutured with perfect safety, no possibility of hernia, no intestine adherent to the scar, and no interintestinal adhesions.

The lesions existing in cystic ovaries never require removal of the organs. Most of the symptoms accompanying them are due to other conditions. In hydrosalpinx we have another condition productive of few symptoms and never threatening life. To open the belly and remove these simple cysts of retention, is to perform an unnecessarily severe operation. To evacuate them through the abdomen and leave them, is to undoubtedly run some risk. Still, evacuation is the proper operation when it is done through the vagina. The same applies to ovarian apoplexy and small broad ligament cysts, to occluded tubes free from pus, and to pelvic adhesions. The great obstacle heretofore to this has been the difficulty inherent in the operations. The incision and posture I recommend render the operations as easy when done through the vagina as through the abdomen.

Coming now to a consideration of acutely inflamed tubes, my task is more difficult. So long as we removed pus tubes and ovaries by celiotomy, our work was eminently proper. Those were the cases which were not checked by the let-alone treatment. But we began to remove the inflamed adnexa in the acute stage, before trying any other measure. There we made a mistake. In many of these cases I checked the infection by an early curettage. But what could be done in the older cases? It is useless to deny to organs so highly vitalized as are the ovaries and tubes, great power of recuperation after infection. The size of the arteries supplying them proves that this possibility exists. It has not heretofore been taken advantage of, because not understood. The acutely inflamed tube becomes rapidly occluded. This is a wise provision of Nature, for no woman could live if a continuous stream of pus was poured into her peritoneum. But this isolating occlusion, while saving life, destroyed the integrity of the affected tube; for the suppuration continuing in the closed tube results in the formation of a pyosalpinx.

Naturally the question is suggested, whether this occlusion would occur if free escape of the tubal nastiness were possible. In other words, the pathological peritonitis is beneficent, in that it saves life, but it is unnecessary where life is not in jeopardy. And where unnecessary, it does not occur. Furthermore, when not present, because relief from the infection is found, the vitality of the affected tube is but little interfered with, and its power of repair is vastly greater than it would otherwise be.

Simply expressed, if I relieve a causative endometritis by curettage, open the cul-de-sac, and open and drain the resultant acutely-inflamed tube, that tube will get well. Women so treated have recovered symptomatically, and their pelves have become free from appreciable lesions. They conceive and go to full term. I can not do this always, because I cannot always accurately measure the extent of damage done. To extirpate an acutely-inflamed tube during a first attack, is simply to deny the possibility of repair. Still we have all seen women stupid enough to refuse operation, and they have got well without it. These few women can be made the many if those forces which bring recovery to the few are appreciated and taken advantage of.

I can safely say that pelvic suppuration can be prevented. The rules governing these operations are laid down under endometritis and salpingitis.

No man nowadays dare assume responsibility for the results following the morphin-poultice-douche treatment of pelvic inflammation. As surely is he responsible for the hysterectomy which will some day be the result of his timidity, as he will be for a death from the neglected disease. It is to teach how to *prevent* suppuration, as well as how to cure it, that I have written this book.

The conditions to which I invariably apply conservative procedures through the vagina are: *hydro-salpinx*, whether unilateral or bilateral; *cystic ovaries; apoplexy of the ovary; occluded tubes; small broad ligament cysts*, single and multilocular; *adherent retropositions*.

The conditions to which I generally apply conservative operations are: *Acute purulent salpingitis; acute puerperal pelvic lymphangitis and peritonitis; pyosalpinx* in young women when seen in first attacks of the inflammation; *recurrent salpingitis*.

I occasionally do a palliative operation in: *Diffuse pelvic suppuration; ovarian abscess*, and in other cases of pelvic suppuration where the general symptoms are too grave to warrant an immediate ablation.

The conditions in which I generally do a radical operation are: *Diffuse pelvic suppuration; genital sclerosis; gonorrheal pyosalpinx* in women over thirty; *relapses after conservative operations; uterine tuberculosis; chronic metritis* with infected ligatures after abdominal operations upon the adnexa; *abdominal sinus* left after celiotomy for adnexal disease; *ectopic gestation* which has ruptured, or unruptured and associated with adnexal disease on the other side; *small bilateral ovarian cystomata*.

Having classified my cases in this way, I may state that

the extraneous circumstances surrounding my patient often compel me to operate in the face of what my judgment indicates would be for her better ultimate interests. To some women the possession of even badly diseased organs is more precious than health; and to others the consciousness that they have lost their special organs is worse than death. These, I know, are sentiments; but I believe them to be held by men also. Castration is an excellent operation for hypertrophy of the prostate, but I am not aware that it is received philosophically by men, nor often allowed. The worst that can be said of conservative operations is that they sometimes fail to relieve. The same is true of all radical work. But when conservatism fails the patient is in no worse state than before, and radical work may still be done.

In those cases, such as pronounced suppuration, where conservatism does not succeed in affording that measure of relief expected, at least the operation removes the patient from the class of emergency operations to one in which the radical operation can be made elective. That much cannot be claimed with reason to attach to the primary radical operation.

Failing to cure by a conservative procedure applied to a pus case, the necessary mutilating operation can later be done in a comparatively clean field, with the kidneys not taxed to eliminate toxins, with the bowel functions restored, with the heart muscle recovered, and altogether with the general condition most propitious to a successful result. No man who has operated in the stage of acute infection with breaking down of the tissues, but will eagerly grasp the opportunity to convert his case into one free from the disagreeable and dangerous elements attaching to the first state. Where conservatism does not succeed in curing, it at least accomplishes that.

CONSERVATIVE OPERATIONS UPON THE INFLAMED ADNEXA UTERI.

Acute Salpingo-oöphoritis.—If the case has progressed too far to be relieved by curettage alone, the efforts of the surgeon should be directed to the prevention of suppuration. Up to recent years two lines of procedure were open to us : either to let the case alone, or else to remove the diseased adnexa. Let us consider a case in its early stages before suppuration has begun. The pelvic peritoneum in its efforts to limit and shut in this infection throws about the adnexa a mass of lymph. This is a beneficent and protective act and is usually effective. Were it not for this isolation of the diseased organs, it is to be presumed that a general and fatal involvement of the peritoneum would result. But at the same time a struggle is going on in the parts inflamed, between the invading germs and the resistant power of the tissues. To overcome the invasion, either a suppurating, destructive process results, or else a connective tissue hyperplasia follows. Either destroys, partially at least, the functions of the organs involved. Is it not possible to check these processes somewhere? It has been determined that removing the causative focus and draining the uterine ends of the tubes and lymph streams by curettage is not sufficient to restore the diseased adnexa to a condition approaching the normal. We must go outside the uterus and drain. Whereas we consider the effusion of lymph and the production of connective tissue essential in the natural process, they are still destructive. Up to a certain stage we can check them and effect a cure. It is in the very earliest stage of the adnexal disease that we can do this.

Operation.—If the uterus has not been previously curetted this is now done (see curettage). If curettage has been done some days before, the uterus is irrigated with boric acid solution. Upon opening the posterior cul-de-sac, serum and lymph flakes escape. The finger is inserted into the cavity behind the uterus, and proceeding toward the lateral pelvic walls all the tender lymph planes are easily severed by the finger. The tubes are freed from their attachments to broad ligament or viscera and gently brought to the vaginal vault for inspection. It is not a difficult matter to open the fimbriated ends with any blunt instrument, the tubes being held by Luer's forceps. A strip of iodoform gauze is inserted into the tube to the uterus. This is left in place until the operation is over. A small amount of fluid may escape from the tubes, clear or cloudy. It is now proper to wipe the pelvis dry. The ovaries are palpated and loosened from adhesions. The operator makes his investigation of the broadest kind. No false attachments between the organs should be overlooked. Every lymph plane should be entered and broken up. Convinced that the tubes are opened and that no organs have been left matted together, the gauze pads are removed, the pelvis is carefully wiped dry, and the strip of gauze in the tube is withdrawn. The uterus is packed with iodoform gauze. Into the opening in the cul-de-sac strips of iodoform gauze are inserted so as to snugly fill the opening. These extend up behind the uterus to the level of the internal os. The uterus and dressings are lifted up into their normal position in the pelvis, and the vagina is packed with gauze. In two days the vaginal and uterine packings are removed and the vagina again packed. The cul-de-sac dressing can usually remain for a week. It is then removed and renewed, sometimes under chloroform. The dressings are renewed about once in five days until the wound closes. The operator seeks to open the lymph streams and tubes so as to cause them to leak. This he would not dare do had he not provided through his gauze a means of escape for the discharges. There no longer being a necessity for locking-in infection, the tissues do not attempt it. The curetting having cut short the source of infection, no fresh supply is furnished. The causative focus in the uterus is removed, and the complications are attacked by evacuation. The question is suggested, Does not lymph form about the gauze in the cul-de-sac? Undoubtedly; but I wish to call attention to the difference between the character of the lymph which forms about an absorbent antiseptic dressing and that which is the exponent of infection. The first is not accompanied by pain, by fever, nor by pus; it is evanescent and produces but few bands of adhesions and these not permanent. Furthermore, it is limited to the cul-de-sac and does not implicate the tubes. Lymph the result of infection is absolutely different. Its production is accompanied by fever, by occlusion of the tube, by thickening of the ovarian capsule, by great pain; and it is permanent or else results in the stoutest kind of adhesions. Moreover, it is extensive in its distribution. The operation is the counterpart of another where the infected focus is cleaned out and the limb above incised to allow of escape of the products of the results of the progressing infection, as in cellular infection of the hand and arm. In very many cases I have done this operation, and never have I failed to check the process. The operation goes a step further than curettage. It is not only conservative, but is curative. To deny it to the woman is to refuse to believe that her most highly vitalized organs have power of repair when aided by incision and drainage. It is absurd to state, as some do, that there is nothing between the let-alone policy of the midwife, and the mutilating operation. From the moment the adnexa are attacked by infection, evacuation and drainage govern us. This operation becomes in the hands of the practitioner the means by which he prevents suppuration, and by applying it early he cures his cases permanently. It certainly takes some courage to come from behind the protection of the hypodermic syringe and thrust oneself into the position of responsibility for the result. Morphin, the poultice, and hot douche but lull the patient into a state of insensibility to her danger. To apply these is to do nothing; to replace them with this operation is to speedily and permanently cure these patients. Not the least attractive attribute of the operation is the ease with which it may be done. It is entirely free from danger.

Chronic Salpingo-oöphoritis.—In case the disease has progressed beyond this first stage of cellular infiltration and there has been a production of pus, the treatment is different. It matters not whether the pus be in the tubes or ovaries. The uterus, unless previously curetted, is cleaned out by curettage and irrigation. The patient is placed in the lithotomy position. Upon opening the cul-de-sac, the operator cautiously works his finger up behind the uterus. When he has reached the fundus, and while doing this he makes firm down-traction by means of blunt bullet-forceps hooked into the cervix, he carefully determines the contour of the ovaries and tubes. When sufficient space has been secured above the diseased adnexa, gauze pads secured by strings are gently inserted above them. If a pus-sac is found low down, it is opened by inserting a closed pair of blunt scissors. As the pus escapes the scissors are opened and withdrawn, thus making a broad rent. A finger is now inserted into the opening, and the whole interior of the cavity is explored. All pouches are entered and the pus evacuated. After the flow of pus ceases, the edges of the sac are grasped with Luer's forceps and held apart, while the operator temporarily packs the sac with iodoform gauze. The pelvis and vagina are wiped dry, and after cleansing his hands the operator seeks possible foci in the other ovary or tube. When found these are similarly treated. After cleaning out all the pus sacs, the field of operation is thoroughly sponged. Under no conditions should the pelvis be irrigated, lest pus be washed up into the higher cavity. It is not advisable to sever the adhesions above the diseased organs. The isolating dome of lymph which usually exists at the pelvic brim is to be left undisturbed. So far the operation has been one of evacuation only. By means of the dressings to be applied the operator seeks the obliteration, by production of connective tissue, of the affected cavity. The gauze pads are removed.

Holding open the pus-sacs, each is filled with iodoform gauze, the ends of which project into the vagina. After this the pelvis itself is tightly packed with the same dressing. The uterus is now packed with gauze. No attempt at replacement is made, but the organs are left in the position in which they are found. The vagina is packed. The uterine packing is removed in two days. At the first general dressing in a week chloroform is given. As the gauze is removed from each pocket, it is renewed before other pieces are taken out. After all packings in the pus sacs are removed and replaced by fresh dressing, the gauze in the pelvis is renewed. Future dressings are made every four days until the openings close. The operation leaves the organs in a damaged state. These women are sometimes cured of all symptoms, but commonly they have some pelvic pain. They menstruate, but are sterile when the adnexa of both sides have been involved. In the course of time the tube or ovary so treated becomes a mere mass of connective tissue. The case assumes the characteristics of genital sclerosis, and the after-treatment is that of sclerosis. Before the sclerosis becomes complete, another infection may set up suppuration again; but where both ovaries and tubes have been treated in this way and have become finally obliterated, I have not seen suppuration occur.

Pus formation is not to be expected in tissues sclerosed by connective tissue. In no sense does this operation resemble the old puncture by means of trocar. When the trocar showed pus, it did not thoroughly evacuate it, and no protection was afforded against future suppuration. If the trocar failed to find pus, it was not evidence that pus did not exist. The trocar puncture was a blind procedure, and the trocar entered all tissues lying in its path. The operation described is safe, thorough, and essentially scientific.

Should a patient so treated, at some subsequent time again become infected, with the production of pus in the pelvis, an immediate evacuation or the radical operation is indicated. The urgency in the indication lies in the fact that the organs no longer have unbroken walls, and hence pus soon tears through into the general pelvic cavity.

Repeated bacteriological examinations have shown that no matter what the cause of the suppuration, after a few dressings the field of operation is sterilized. No pyogenic cocci are found, but the colon bacillus is very constantly present. All that is apparently necessary to induce the presence of this germ is any traumatism inflicted upon the vagina or retro-uterine structures.

Chronic Lesions.—*Hydrosalpinx.*—These simple cysts of retention have heretofore been treated by removal through the abdomen. As early as 1891 I became convinced that they were inocuous, but up to five years ago had not attempted their treatment through the vagina. The fluid they contain is very generally sterile serum, and its evacuation into the pelvis produces no more reaction than the presence of peritoneal fluid. No tube the seat of hydrosalpinx should ever be sacrificed.

Operation.—The cul-de-sac is opened after curettage of the uterus. The uterus is held down by the traction forceps, and the affected tube is easily freed. When it is exposed, with a blunt pair of scissors it is incised for an inch along its upper border, beginning at the fimbriated end. The fluid is caught with gauze as it escapes, and the pelvis is wiped dry. But little oozing takes place unless many adhesions have been severed. Unless an indication exists for draining the pelvis, the incision in the vagina is sutured by a continuous suture of chromic catgut, and the uterus is packed with iodoform gauze. It is then replaced and the vagina tamponed with iodoform gauze. The uterine packing is removed in two days and the vaginal dressing renewed. The patient is allowed out of bed on the tenth day, the vagina being kept tamponed until it is entirely healed and until the sutures have become absorbed. This latter occurs in about two weeks. Whenever the operator is in doubt regarding the propriety of closing the vaginal incision, it may be kept open by a packing of gauze. The convalescence is afebrile and the recovery complete. Sometimes the walls of an old hydrosalpinx are thick and ooze when incised. Beginning at the fimbriated end of one side of the cut a running suture of fine catgut is taken down one side of the cut to the angle of the incision in the tube, and then along the other side to the fimbriated end. In this way the peritoneum of the incision is folded over to the lining membrane of the tube (Fig. 55). Oozing is thus checked

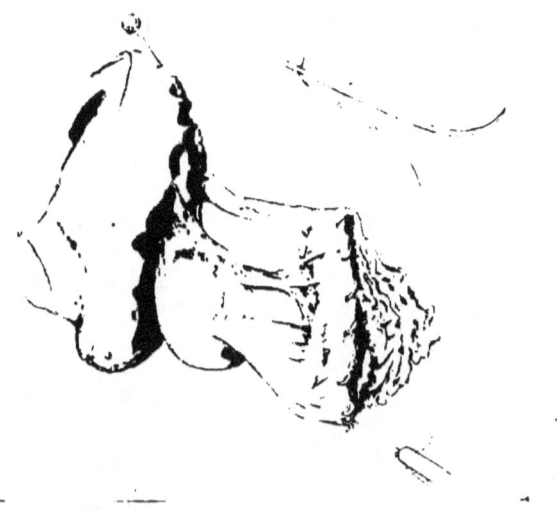

FIG. 55.—Salpingostomy. The occluded tube has been incised along its upper border, and a running suture is being taken so as to bring the mucous lining to the serous covering of the tube, and in this way maintain the tube's patency.

and closure of the tube prevented, "salpingostomy." The cul-de-sac is not closed in such a case, but is packed lightly with gauze. In all hydrosalpinx cases the uterus if retroverted is replaced.

Cystic Ovaries.—These should never be removed. They produce but little pain and cause only mild reflex symptoms. The uterus is curetted and the cul-de-sac opened. The ovary is freed from false adhesions and brought down into the vagina; where it is held by bullet forceps. One by one the cysts are stabbed with a tenotomy knife. Sometimes a cyst is met with of large size, even one inch in diameter. It is to be evacuated, its edges are trimmed, and the membrane which usually lines it is peeled out. The cut edges are then brought together by a running suture of fine chromic catgut.

After all cysts are emptied the ovary will be found much shrunk. (See Fig. 32.) It is returned to the pelvis, and the vaginal incision closed or packed as indicated. The uterus is packed with iodoform gauze and replaced by a vaginal tamponade of the same material. The after-treatment is the same as for hydrosalpinx.

Ovarian Apoplexy.—The uterus is curetted and the cul-de-sac opened. The ovary is freed and pulled into the vagina with bullet forceps. While held there the blood cyst is incised, and the contained clot evacuated. The edges of the cyst are trimmed with scissors, and then the lining of the sac is peeled out with Luer's or other suitable forceps. After this is done it will be found that the ovary is much reduced in size. If the resultant cavity is large, I trim it so as to form two thick flaps which I suture readily with a running suture of fine silk or chromic catgut. I prefer the latter. If the cavity is small, I leave it open and do not bother to sew it. It may ooze a little, but no more than would a ruptured Graafian follicle.

I usually leave the vagina open in these cases and pack with gauze, because these cysts are prone to be of remote septic origin. The uterus is packed with gauze as also is the vagina. The after-treatment is similar to that of other non-suppurating cases.

Adherent Ovaries and Occluded Tubes.— The uterus is curetted and the cul-de-sac opened. All adhesions are broken with the fingers or else held up with a blunt hook and severed with scissors. The occluded tube is drawn into the vagina and incised along its superior border for an inch from its fimbriated end.

The fimbriæ are teased apart with forceps. While holding apart the edges of the V-shaped cut, a running suture of fine catgut is taken from the upper border of the fimbriæ down to the angle and up to the fimbriæ of the lower flap. (See Fig. 55.) This suture is so applied as to unite the peritoneal surface with the lining of the tube, and is used for the purpose of preventing closure of the tube. The uterus is packed, and the opening in the cul-de-sac filled with gauze which reaches just within the cut edges. The vagina is packed with gauze. The usual after-treatment is employed.

Broad Ligament Cysts.—When these are purely pelvic, whether single or multiple, they can be treated through the vagina. When they reach up to the pelvic brim they should be removed by laparotomy.

The uterus is curetted and the cul-de-sac opened. At once the smooth thin-walled cyst is felt. It has no pedicle; therefore, the uterus is held up with a trowel while the posterior vaginal wall is drawn down. Gauze pads are inserted above the cyst and the intestines kept up. With blunt scissors the cyst is split open and emptied. A portion of the flaccid posterior walls of the cyst is torn away with Luer's forceps. The pelvis is wiped dry, and the pads removed. No bleeding of consequence results. The uterus is packed with gauze, the cul-de-sac is filled with the same material which reaches up to lower margin of the cyst cavity, and the vagina is packed. The usual after-treatment is employed.

The After-Treatment in Non-Purulent Cases.—In two days the vaginal dressings are removed and the uterine packing withdrawn. The vagina is again packed. From eight to ten days after the operation the patient is placed in Sim's position and the cul-de-sac dressing taken out. In doing this the uterus must be supported by the trowel. Fresh dressing is inserted and the vagina again packed.

The second dressing is made in a week more, after which the patient is allowed up. The cul-de-sac is kept packed until closed.

If the vaginal incision has been sutured, the sutures are removed in two weeks, and the vagina kept packed until the scar is stout.

I do not give douches until the wound is healed, and forbid intercourse for six weeks after the patient is discharged.

PREPARATION OF PATIENT FOR A VAGINAL SECTION.

General.—The presence of nephritis, of cardiac disease, or of phthisis is no bar to the operation. Where patients have influenza I prefer waiting for a few days until this subsides, lest the narcosis excite a bronchopneumonia. Five days before the operation the patient is given a calomel purge. I prefer triturates of calomel each of gr. ¼, given at 7, 8, 9, and 10 P.M., to be followed next morning by a saline purge, like Seidlitz powder. The diet is general and includes everything but the more indigestible foods and luxuries. I exclude everything fried, whether vegetable or flesh; stimulants are withdrawn, and narcotics, if previously used, are not allowed. The patient is made to lie down most of the time, reading periodicals, seeing few friends, and altogether assuming a semi-invalid state. She is encouraged to drink large quantities of water. Each night she is given a high enema of normal salt solution, of two pints. This she is encouraged to retain. The object is to charge the tissues with fluid. This has been shown not only to actually increase the amount of urine, but also to facilitate the elimination of urea. I have the urine analyzed for sugar, albumen, and per cent. of urea, the total for twenty-four hours being carefully measured. If fever has existed before this treatment is instituted, it usually diminishes, and if there is albumen in the urine this decreases. I strive to get the emunctories cleansed out and at the same time store up an excess of fluid for the day following the operation, when the kidneys take away at least from half a quart to a quart of urine and no fluids are ingested. The shock, both surgical from hemorrhage, and vasomotor from traumatism to these important pelvic structures, is much diminished.

Local.—Two days before operating I prepare the patient. The pubes and vulva are shaved, the abdomen is covered by a wet dressing of ½ per cent. lysol solution, and the vagina is packed with gauze wet in bichlorid solution $\frac{1}{1000}$. These dressings are changed twice more before the operation. The day before operating I give meat once, potatoes, bacon, eggs, tea, soup, as much as needed. All the time the patient is instructed to drink two quarts of water a day. I do not like milk. In the first place it has poor food value for an adult, and its digestion results in the formation of "bullets" in the bowels. Furthermore, the intestinal gases are increased by it. For the same reason I do not use koumyss.

After the first purge of calomel and salts, it is rarely necessary to use another laxative; but if needed, one pil. rhei. comp. may be given two days before operating.

The day I operate I give no food or drink after midnight, if the narcosis is to be in the forenoon. If I operate in the afternoon, I give coffee and toast for breakfast and a pint of water at 11 A. M. I do not give stimulants either before or during the operation. Certain very desperate cases are met with: those with nephritis and prolonged suppuration. In such cases I perform transfusion into the elbow vein, using c. p. normal salt solution, and introducing as much as sixteen to sixty ounces at the time of operating. To old drunkards and to women with fatty hearts, I sometimes give a hypodermic of strychnia, gr. $\frac{1}{30}$, before beginning the narcotic. But I do not do this often.

Cleansing the Patient.—The position is that for lithotomy, with the coccyx hanging over the table. The thighs and vulva are scrubbed with lysol solution 2 per cent. The packing is withdrawn from the vagina, and the latter is scrubbed with 1 per cent. lysol solution, using for this purpose a long brush (jeweller's,

(Fig. 56). The perineum is depressed and the brush moved up and down and rotated within the vagina while an assistant pours the solution into the vagina. After

FIG. 56.—Brush for scrubbing the vagina.

using the lysol the external parts and vagina are scrubbed gently with Thiersch solution. The legs and all parts of the field of operation are covered by sterilized towels or stockings. The operator then proceeds.

VAGINAL ABLATION.

General Considerations.—The vaginal mucosa and peritoneum only are severed in vaginal hysterectomy; whereas, in laparotomy, the skin, fat, fascia, muscle and peritoneum are cut. In vaginal hysterectomy no vessels are cut by the incisions which require ligation, but many small arterial trunks must often be secured in making an abdominal wound. It is not necessary to sever the peritoneum in performing vaginal ablation, for sufficient space may be secured without that. It is necessary to dissect the uterus from the bladder in both vaginal and abdominal ablation, but in the former the advantage is present of having the cervix as a guide. The uterus and adnexa to be removed are not masked by the viscera which lie above them when vaginal ablation is done.

Separation of Adhesions.—It is usually necessary to work through a mass of adherent intestines before the organs we seek are seen in laparotomy, while the work in vaginal ablation proceeds below the matted intestines which lie above the uterus. This attribute of the vaginal operation is worthy of a moment's discussion. We find two kinds of adhesions: Those which have formed between the various coils of intestine, the *inter-intestinal;* and those which exist between the organs to be sacrificed and the intestines. We do not disturb the inter-intestinal adhesions at all when we perform vaginal ablation. The point may be raised that it is advisable to dissect the intestines free; but the weight of opinion is in support of the belief that when this is done, not only do the adhesions re-form, but that the secondary union is more general when we deal with pus cases, such as are under discussion. Certainly the experience is exceptional with all of us to find upon making a second laparotomy that there are not evidences throughout the track of the operation of a pretty general infection resulting from the first section. Furthermore, in performing even a primary section in these pus cases, the inter-intestinal adhesions are so firm that breaches are made in the intestinal walls, often requiring suture. Regarding the adhesions between the uterus and the organs to be removed, whatever raw surfaces are made upon the intestines in the vaginal operation, remain turned down toward the point best adapted for drainage; whereas, in the abdominal operation all the raw surfaces are dragged up above the pelvic brim, with a possibility of infecting all points from which manipulation has removed the endothelium.

In the vaginal operation, only those false unions are severed which bind the diseased organs to be removed, and these are much less important than those which exist at and above the pelvic brim.

We are forced to the conclusion that any operation which, other attributes being about equal, will furnish an escape from the tedious dissection often incident to a laparotomy in pus cases, will bestow the greatest immunity from one most disagreeable sequela of laparotomy. In abdominal section very often a grave intra-peritoneal operation has been done before the organs sought for are even seen. Usually the uterus and adnexa are removed *per vaginam* without a knuckle of movable intestine being seen.

Direction of Effort in the Enucleation.—In laparotomy the operation proceeds through an incision which is expected to heal by first intention and through a mass of adherent intestines. The infected organs are dragged up between the raw surfaces left after separating the adherent intestines and between the margins of the abdominal incision. The fingers whether naked or gloved repeatedly take the same path, and no hand which has been engaged in liberating and removing pus foci can be insured as clean. In laparotomy the organs removed are dragged from their pelvic attachments through the lower part of the abdomen. In vaginal ablation the direction of the effort is in the direction of drainage at the lowest part of the peritoneal pouch. The pelvic filth remains pelvic and is never led into the abdomen. It does not pass by tissues which are to be sutured, and does not infect areas of intestine from which the endothelium has been removed by manipulation.

Hemostasis.—In laparotomy this is by means of ligatures which must be absorbed; certainly those upon the ovarian vessels are cut short and left in. These ligatures are so frequently infected, being placed in an infected field, that they are often sources of trouble although isolated in a mass of lymph. All the problems embraced in a consideration of the choice of ligature material, its preparation and its fate, are factors when the operation is done through the abdomen. They are not considered in the vaginal operation.

Drainage.—In laparotomy this must sometimes be employed, particularly in cases of streptococcus infection, diffuse suppuration, and where tubo-rectal fistulæ exist. As a result the isolation of the area drained is effected by a matted mass of lymph thrown out by the intestines, and a breach is left in the abdominal scar. Besides, the pelvic filth is drained through the normal abdominal cavity, and is up-hill. In vaginal ablation the drainage is always used; it is at the lowest part of the pelvic cavity; the intestines do not become adherent to the drain or area drained, the pelvic filth remains pelvic, and drainage is down-hill.

Drainage after laparotomy, though not often used now-a-days, infects the entire area adjacent to the drain from the pelvic floor to the abdominal skin. Drainage after vaginal ablation passes for not over an inch through the lowest part of the pelvic peritoneum, and most of it is through the vaginal tube which is particularly adapted to carry off the material drained away without absorbing any. The infected drainage space after laparotomy remains for a large part an abdominal complication, and for weeks.

After vaginal ablation, the drainage track is in a few hours made extraperitoneal by the union of bladder to rectum.

Sutures.—These are not used in vaginal ablation. So important a matter is the method by which the abdominal wound should be closed that there are about as many varieties as there are operators. Shall the wound be closed by buried catgut, buried kangaroo tendon, or buried silver wire? Shall the wound be united by suturing in tiers or through-and-through suturing, or shall the fat be left open? Shall the sutures be applied as interrupted or mattress or continuous sutures?

Hernia.—The percentage of hernias after laparotomy is not known, but there are many of them. They are not known to follow the vaginal ablation by forceps. The intra-abdominal effort is almost wholly borne above the symphysis, while the vaginal vault is protected from this force by the posture of the body and the sacral promontory.

Accidents.—In abdominal hysterectomy the bowel must sometimes be sutured; the ureters have been cut; abdominal fistulæ are known to exist, and ligatures have worked their way into the bladder. After vaginal ablation intestinal suture and resection must be exceedingly rare procedures; in the few cases in which the bladder has been wounded the rents closed without suture, unless made by the veriest tyro; no wandering ligatures are heard of, and no abdominal fistulæ are found.

Instruments.—In laparotomy knives, scissors, needles,

sutures, ligatures, needle-holders, etc. In vaginal ablation no needles, no sutures and no ligatures. Much less complicated is the preparation for vaginal section.

Narcosis and Time.—Abdominal hysterectomy necessitates an abdominal section and a hysterectomy. Vaginal ablation is a hysterectomy only, without the abdominal section. Few men can perform a *finished* abdominal hysterectomy in less than three-quarters of an hour in *pus cases*. Twenty minutes only need be consumed in vaginal ablation. In order to secure relaxation of the abdominal muscles profound narcosis is necessary in laparotomy. With vaginal ablation the narcosis is incomplete and short, and chloroform again becomes the preferable anesthetic.

Convalescence.—No man who has seen a number of similar cases treated by the two methods but will decide that the ability to turn over in two days, the assumption of regular diet in four days, the regularity of the bowels from the first, the absence of nausea and vomiting, the early getting-up, make the convalescence from vaginal ablation much less disagreeable than from laparotomy.

Results.—No case of mine has died either from the operation or from complications. There are no fecal fistulæ to report, no sinuses, no vesicovaginal fistulæ, and no hernias. There have been no cases of phlebitis and no intestinal obstructions. The vagina has in no case been shortened, and intercourse is painless.

These are the reasons why I perform vaginal ablation in pus cases.

Having stated my reasons for preferring the vaginal route, I may properly mention what cases I exclude from the list of those to which I apply this method. Any important bowel complication above the pelvic brim must be treated through the abdomen. Whenever a suppurating ovary or tube communicates with a purulent vermiform appendix, and whenever a pus focus in the adnexa opens into the *small* intestine, or the large intestine above the pelvic brim, the bowel lesion so far overshadows the pelvic disease that the case must be viewed from the abdominal side, for the delicate suturing of the intestine can not be done through the vagina. The question is natural, Can these facts be determined by vaginal section? I have not found any difficulty in doing so since the perfected technic has been adopted.

Whenever hysterectomy is indicated upon a puerperal uterus, the vessels are so large that by hemisection too much blood is lost, hence the uterus must be removed *en masse;* and the tissues are so friable that the requisite down-traction is impossible. Vaginal ablation of a puerperal uterus is truly a deplorable operation and one which should never be done.

But there are a number of cases which when treated by laparotomy give a very high rate of mortality—I refer to cases of diffuse suppuration. When approached by the vagina these are as successfully handled as any others.

Posture.—The uterus can be removed *per vaginam* with the patient in the lithotomy posture throughout the entire manœuvre; but the operator will find that he can proceed with greater comfort to himself and safety to his patient if he employs a table which enables him to secure the Trendelenburg posture. The one I have devised can be employed in all gynecological work and is particularly useful in the vaginal operations; but a suitable table may be improvised by sawing off two of the legs of a stout kitchen table so that the incline of the table will be 60°. To retain the patient in the lithotomy posture I employ Ott's or Clover's crutch; but a sheet passing over her shoulders and tied to the legs will answer.

Having a table which enables him to secure the Trendelenburg posture at any moment, the operator can avoid all those accidents which accompany improperly applied and imperfectly protected forceps. If he so desires, he can operate in a pelvis which is entirely free from abdominal viscera (Figs. 57, 58, 59).

Operation.—I operate standing. The field of operation is cleansed, and the uterus is curetted and swabbed out, but is not packed. All instruments used in the

FIG. 57.—Operating table folded for transportation.

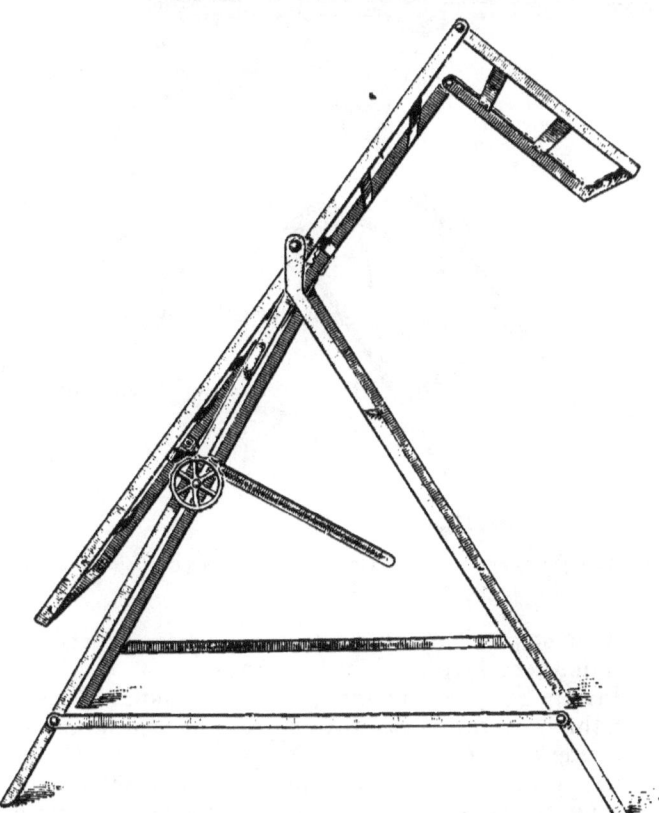

FIG. 58.—The table in exaggerated Trendelenburg position for laparotomy.

curettage are laid aside, and the operator again washes his hands.

The Incisions.—I always attempt and rarely fail in inflammatory cases to enter the posterior cul-de-sac as

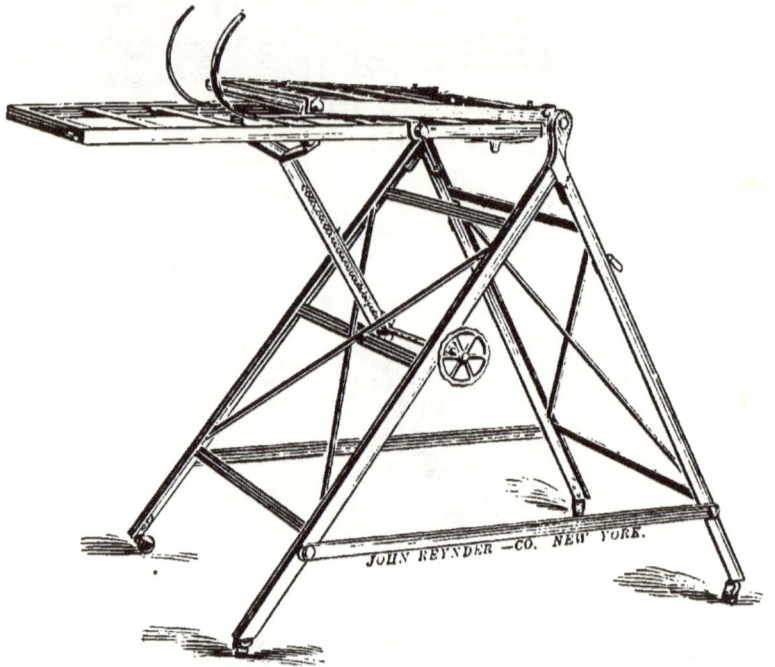

FIG. 59.—The author's table arranged for vaginal operations in the lithotomy posture.

the first step. This is the true exploratory part of my operation (see Exploration, page 136).

Having become convinced that an ablation is necessary, the operator proceeds to spread the vaginal incision from side to side (Fig. 52). The posterior incision having been completed a gauze pad is introduced into the opening to catch fluids. The anterior incision is next made. I introduce into the uterus a pair of my intra-uterine traction forceps, and spread them until a firm grasp is secured upon the organ (Fig. 61).

The cervicovesical fold is accurately determined, and,

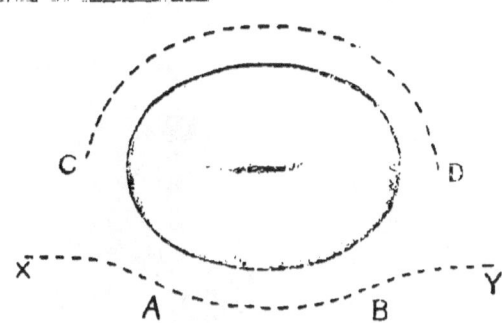

FIG. 60.—The lines of incision in the vaginal operations. A-B, the extent of the incision into the posterior cul-de-sac for the purpose of severing adhesions. X-Y, the incision for evacuating pus, in puerperal fever and in hysterectomy. C-D, The anterior incision for dissecting off the bladder.

cutting against the cervix, the latter is circled to within an eighth of an inch of the posterior cut. Thus a narrow strip of vaginal mucosa is left upon each side. I do

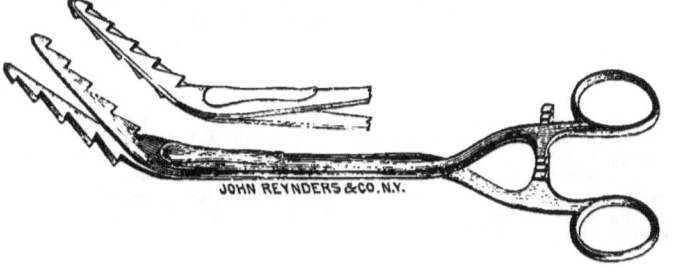

FIG. 61.—The author's intra-uterine traction forceps.

not make this anterior incision near the external os. I wish to cut above the very dense tissues about the external os and yet to leave abundance of vagina. If the dissection is made near the os, bilateral space is secured with difficulty, for the incision will be surrounded by a ring of inelastic tissue (Fig. 62). In other words, the anterior incision should be made in vaginal tissue and not in cervical. So soon as the scissors have cut through the vaginal mucosa, they are closed and laid sideways upon

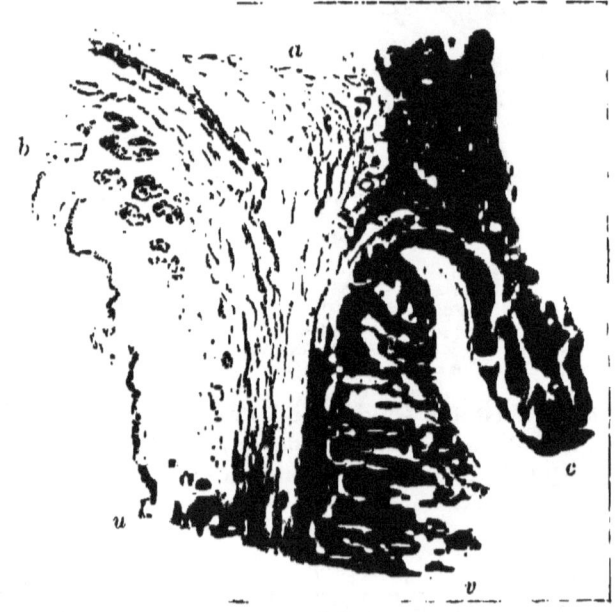

FIG. 62.—Shows microscopic section (vertical) of fetal bladder, urethra, anterior vaginal wall, and anterior lip of cervix. *a* lies above loose tissue, between bladder and cervix and vagina; *b*, bladder; *u*, urethra; *v*, vagina; *c*, cervix uteri. X marks the point at which the anterior incision is made, so that the dissection may proceed through the loose tissue between cervix and bladder (Hart).

their edge in the cut. Bearing down hard upon the cervix, the tissues are shoved up for a short distance or until the looser tissues are reached. The closed scissors used in this way act as does a periosteum elevator. After the dissection has proceeded upon the anterior face

VAGINAL ABLATION.

of the cervix for about half an inch, a short retractor is inserted into the wound and the bladder held up. Upon wiping the wound dry a few bands of connective tissue and muscular fiber may be noticed extending from the sides of the incision toward the center and angle of the denudation. These are snipped with the scissors. After this all attempts to enter the anterior peritoneal pouch are made with one finger. Holding the uterus firmly with the intra-uterine traction, the vesico-uterine tissues are pushed up. The operator does this by bearing hard

FIG. 63.—Transverse section of right half of uterus at level of the internal os. *c*, bladder; *a*, uterus; *b*, parametric tissue; X, uterine vessels (Hart).

down upon the uterus with the index finger and literally *rubbing* the bladder tissues from the uterus. This is done not with the nail, but with the palmar surface of the finger. It is in this bladder dissection that the great value of my forceps is seen. With them the uterus can be rotated so as to differentiate the loose pericervical tissues from the uterine; and in stripping the bladder from the uterus they furnish a most admirable point of counter-pressure. They give the operator a fixed body to work against and not a movable one. I have never found it necessary to sever the peritoneum with instruments. The finger, whenever it can reach the fundus anteriorly, will easily penetrate; and in cases where the peritoneum is attached high on the uterus, the peritoneum should not be blindly opened until the uterus can be pulled down after hemisection. Having entered the anterior fornix or made the dissection as high as the finger will reach, the bladder is separated from the uterus to the sides. The anatomical fact must here be noted that the width of the bladder is greater than that of the uterus, and that the organ extends laterally upon the broad ligaments. The operator sticks to the middle line in separating the bladder and makes the lateral separation by moving the finger, laid flat upon the uterus, from side to side. The uterine vessels at the sides can be felt pulsating, and the dissection should not be carried beyond their level (Fig. 63). If the operator is rough he can very easily rupture the uterine vessels. So far there has been but little bleeding. The azygos artery on the posterior vaginal wall has been severed in opening the cul-de-sac, and temporarily clamped if prominent. The small vessels from the uterine arteries which enter the cervix give some trouble if wounded. They anastomose freely with the vesical arteries. I do not pay attention to them until I am ready to clamp the uterines. The operation has progressed to the point where the uterus is free from its attachments to the bladder and posterior vaginal wall. I have termed this the first stage; for it is done in all cases, be the further manœuvres what they may. In making these incisions and separating the bladder, what is the position of the ureters? At the point where the uterine artery springs from the internal iliac, the ureter lies at least a quarter of an inch below the artery. As the artery abruptly crosses the pelvis to the side of the uterus it passes across the ureter. This point of crossing is always at least an inch from the normal cervix, and is where the broad ligament spreads out for its attachment to the side of the pelvis (Fig. 64). After this the ureter and uterine artery are never in relation. The ureter sweeps in a graceful curve to the bladder and is *in front* of the uterine artery. The uterine artery does not curl around the ureter, as pictured by Bourgery and Jacob. From the time the ureter crosses the pelvic brim, it begins to sink

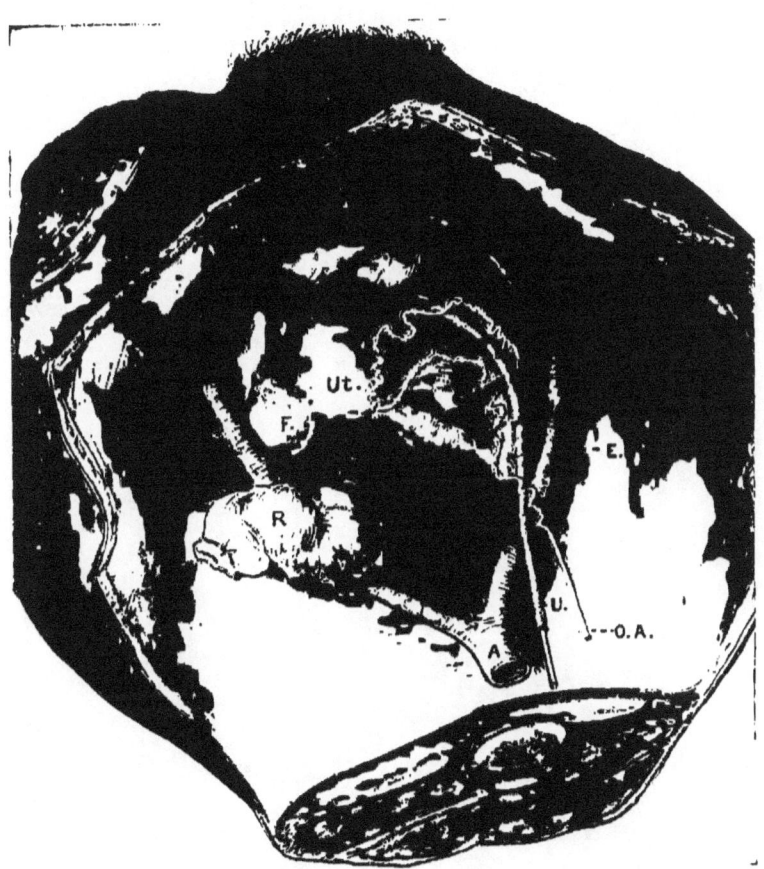

FIG. 64.—A., abdominal aorta; I. I., internal iliac artery; E. I., external iliac artery; O. A., ovarian artery; U. A., uterine artery; R., rectum; F., a fibroid nodule springing from the fundus; Ut., uterus; O., over the right ovary which is adherent to the posterior surface of the broad ligament. Above O is a right hydrosalpinx; U., ureter into which has been introduced a probe; B, bladder; S. V., superior vesical artery. The pelvis is somewhat tilted to the right to show the relations. The peritoneum has not been removed, but the course of the vessels has been shown by painting over their course beneath the peritoneum.

VAGINAL ABLATION.

below the internal iliac artery; and when the uterine artery is reached, the ureter is easily a quarter inch below the uterine. The ureter proceeds anteriorly to the bladder, while the uterine artery crosses the pelvis to the cervix. Upon separating the bladder from the uterus and lifting it up, the ureters are swung outward and further upward; and pulling the uterus down and toward the

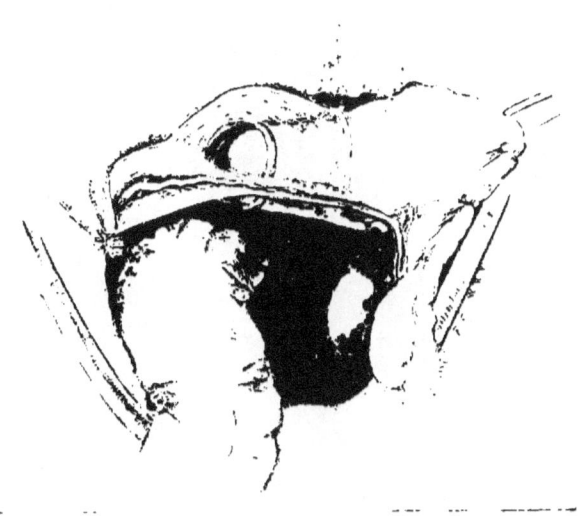

FIG. 65.—A photograph of a suprapubic hysterectomy, introduced for the purpose of demonstrating the position of the ureter, which may be seen crossing beneath the uterine artery at the outer margin of the split broad ligament.

sacrum while lifting the bladder still further moves the uterine artery to a deeper and more posterior position. When the bladder is separated and held up, and the uterus pulled down, the ureters and uterine arteries are further apart than they were before the operation. But if the bladder is not separated and lifted, down-traction upon the uterus decreases the angle of divergence between the artery and ureter, and they may be made to touch for the outer half of the artery and up to a half inch of the cervix. Repeated dissections show this.

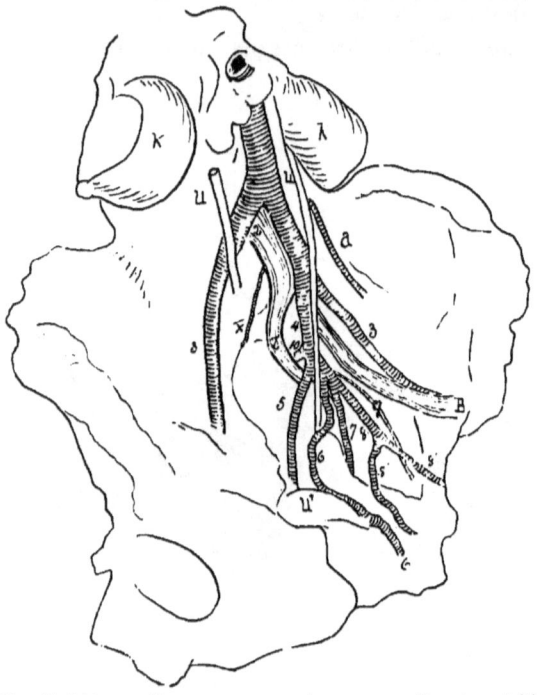

FIG. 66.—K, kidney; U, ureter; a, ovarian artery; B, external iliac vein; Ut, uterus; 1, abdominal aorta; 2, vena cava inferior, on each side of which is a common iliac artery; 2', internal iliac vein; X, middle sacral artery; 3, external iliac artery; 4, internal iliac artery; 5, internal pudic artery; 6, uterine artery; 6', point where uterine artery joins the severed ovarian artery; 7, vaginal artery; 8, 8', superior vesical artery; 9, obturator artery; 10, common origin of the gluteal and sciatic arteries.

The ureters can not be wounded by any force applied at the sides of the uterus, provided such force does not tend to draw the cervix and bladder together, as, for instance, an improperly applied ligature does.

FIG. 66.*—Surgical Anatomy of the Internal Iliac.

I leave a narrow strip of vaginal mucous membrane upon each side and between my anterior and posterior incisions for two reasons. When I apply the forceps to the uterine vessels, this strip of tissue prevents tearing off the forceps during future manipulations. Furthermore, I have thought this prevented to some extent sagging down of the vagina after completion of the process of healing, inasmuch as the vaginal vault and the bases of the broad ligament are one.

I have never found that my incisions gave me less room than Ségond's. This surgeon circles the cervix entirely, and makes upon each side a cut at the base of the broad ligament. I never find it necessary to incise the perineum to gain space. Could I not perform the operation without this, I would always do laparotomy. One attractive feature about this vaginal operation is absence of traumatism to normal structures. This is lost when the perineum is incised.

When Ségond has separated the bladder he has two flaps, and his first pair of forceps do not grasp vaginal mucous membrane at all. Ségond contends that his incision enables him the better to avoid the ureters. In one way it does, inasmuch as there is a greater separation of the anterior (bladder and ureters) segment from the posterior (uterus), not in the middle, but at the sides of the uterus. I have used both incisions and prefer the one illustrated, for the reasons stated.

In certain cases the bladder is attached so high up on the anterior surface of the uterus that the operator cannot reach the anterior peritoneum with his finger. He should then make his dissection as high as he can, and withdraw the intra-uterine traction forceps. In order to enter the peritoneum it is necessary for him to pull down the anterior surface of the uterus. In order to do this, he grasps each side of the cervix with bullet forceps, and splits the anterior lip of the cervix in the middle line to a little above the level of the internal os. (See Hemisection.) Upon rotating the bullet-forceps outward the cervical canal will flare out, and a portion of the anterior uterine wall will come down. This is cut with scissors in the middle line. While making this anterior median section of the uterus, the bladder should be held up by a narrow retractor, and as each successive portion of the anterior wall of the uterus comes into view, it is grasped by traction forceps. After a time, at the upper angle of his incision, the operator will see the smooth peritoneal covering of the uterus. He has, perhaps unconsciously, entered the anterior peritoneal pouch by holding up the bladder and progressively splitting the anterior face of the uterus.

It is well after entering the peritoneum posteriorly and anteriorly to make a careful digital exploration of the pelvis. Now is the time for the operator to obtain an accurate knowledge of the regional anatomy of the particular pelvis he is dealing with. This completes the first stage of the operation, and the procedure is employed in all cases.

ABLATION EN MASSE.

Freeing the Adnexa.—If this can possibly be done before the application of the forceps, it should be, for forceps take up room. The gauze pad in the cul-de-sac is removed. Still pulling the uterus down, the operator inserts his finger into the pouch of Douglas. Taking the posterior surface of the uterus as a guide, he enters the finger to the level of the tubes. After one tube is found, attempts are made to free its attachments at the fimbriated end. In doing this the effort is made to *push* up the tube and ovary. The operator is working from behind the broad ligament. If the adnexa are attached low, they can readily be freed. If they are high at the pelvic brim, the effort to release them is made in front of the uterus, the fingers being between the bladder and uterus. Here the uterus is again the fixed guide. In working from in front of the uterus the operator seeks to free the adherent organs by getting his finger outside of them and separating the attachments toward the cornu. The action is very similar to that used in like cases when laparotomy is done. Having released the adnexa on one side, those of the other are released. Too much emphasis can not be laid upon the importance of persisting in efforts to release the adnexa from inflammatory union to other organs before putting on any forceps. No vessels of importance have so far been severed; the narcosis is not profound, and the patient is in no sort of danger. The operator need not be embarrassed if he finds the adnexa firmly adherent, but must persist in his efforts to free them both by working from behind and from in front of the uterus. It is well to have a firm grasp upon the cervix with blunt traction forceps or the intra-uterine traction forceps, and to work with all specula removed. As few instruments as possible should be in the vagina. While drawing down the uterus with the left hand and manipulating the adnexa, the assistant will render material aid by pressing down from above the pubes. When he has released the adnexa, the operator withdraws his hand and introduces the anterior and posterior retractors (Fig. 67).

Taking in his right hand a pair of hysterectomy forceps, the operator introduces one blade into the anterior incision, to the left of the cervix, and the other blade into the posterior incision. The forceps is crowded, still open, hard up alongside the cervix, and when in position the operator carefully inspects and feels each blade to see that no intestine or omentum is caught. The forceps is then locked. A forceps is similarly applied upon the right side. It will now be seen that all bleeding about the cervix has ceased. Into the posterior cul-de-sac a gauze pad is inserted to hold up the intestines; and the tissues upon each side of the cervix between the cervix and the two forceps are cut with scissors almost to the points of the forceps. The intra-uterine traction forceps or a male sound is now used to antevert the uterus. As high on the anterior face of the uterus as he can see, the operator takes a firm grasp of the uterus with toothed forceps and withdraws the intra-uterine traction (Fig. 68). He shoves the cervix upward while he pulls down on the body of the uterus until he can see more of it, and again takes a good grasp near the fundus. He can now draw the fundus forward beneath the bladder until the cornua appear. While supporting the fundus in this way he inserts his fingers above the uterus, and seizes the right adnexa.

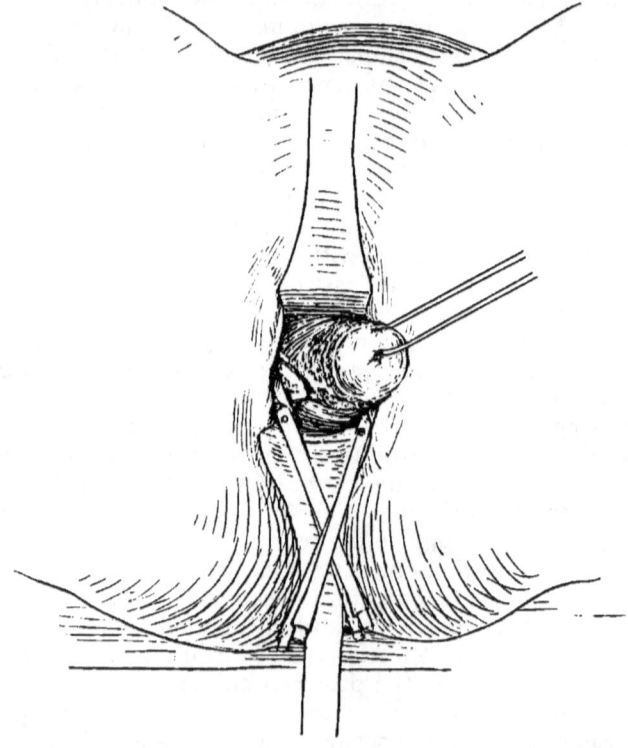

FIG. 67.—The cul-de-sac has been opened and the bladder dissected from the uterus. The uterine arteries are grasped by forceps and the cervix has been dissected from the lateral stumps (photograph of operation).

Either with his fingers alone, or assisted by Luer's forceps (Fig. 69), he drags the right adnexa in front of the uterus and applies forceps to the right ovarian artery outside the right ovary. This forceps is applied from above, and the operator can guide the anterior blade with his index finger, and the posterior blade with his middle finger, so that there is no danger of catching any intestine. It is well to withdraw the gauze pad before applying this forceps lest it be caught in the forceps. When he feels that this forceps laps the one on the right uterine artery,

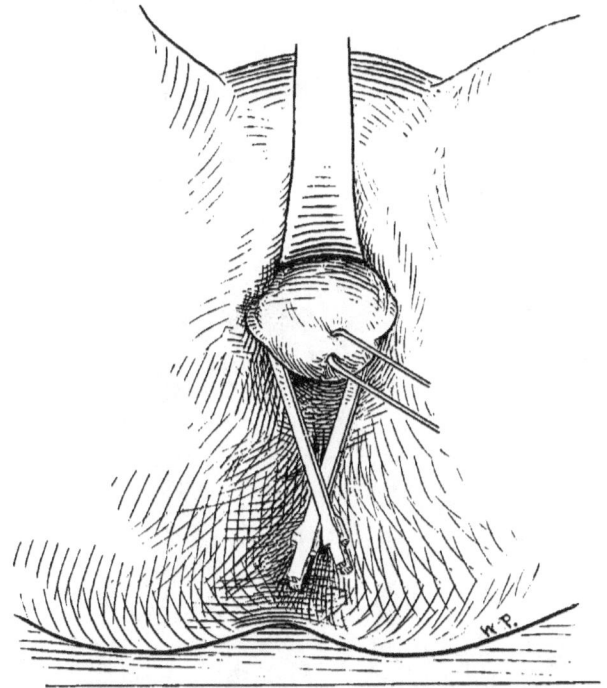

FIG. 68.—The cervix has been shoved up so as to permit the operator to drag the fundus out beneath the bladder. Both cornua uteri are shown with the attached tubes (photograph of operation).

he clamps it. In isolating the right adnexa and applying this forceps, if retractors are used and are in the way, they should be withdrawn. The upper forceps grasps the round ligament as well as the broad ligament. The uterus is now cut loose upon the right. At once it swings out of the pelvis so that its posterior face is forward, and it becomes an easy matter to bring forward the left adnexa and secure the left ovarian artery outside the ovary. The uterus can now be cut away (Fig. 70).

The specula are next introduced. Holding the bladder

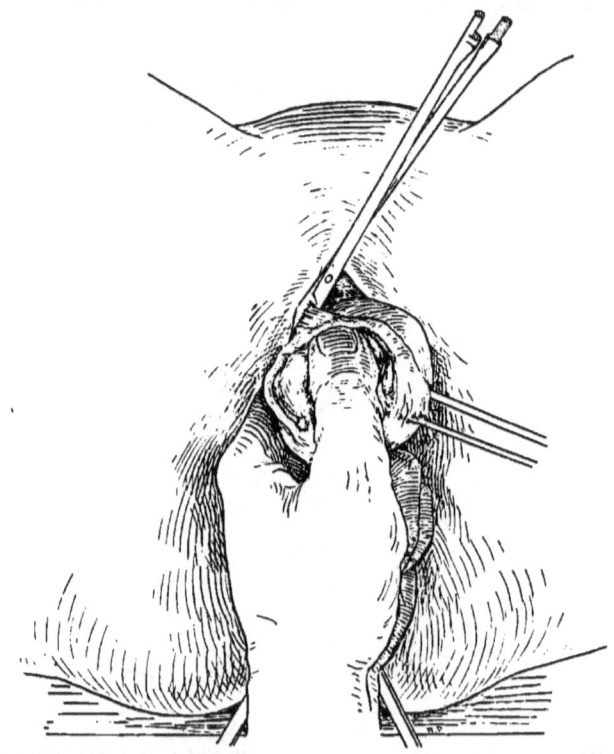

FIG. 69.—After delivering the fundus the entire uterine body is pulled to the operator's right, in order that the right adnexa may be seized. The operator's thumb rests on the ovary, while his two first fingers grasp the corpus uteri. The forceps are being applied to the right ovarian artery. Notice the absence of retractors (photograph of operation).

up and depressing the perineum and posterior wall, the operator introduces a gauze pad into the pelvis and pushes the intestines away from the stumps secured by his forceps, so that he may make a careful inspection of the stumps and see if any bleeding is going on. If the adnexa have been thoroughly freed before extirpation is attempted, it will be seen upon completion of the opera-

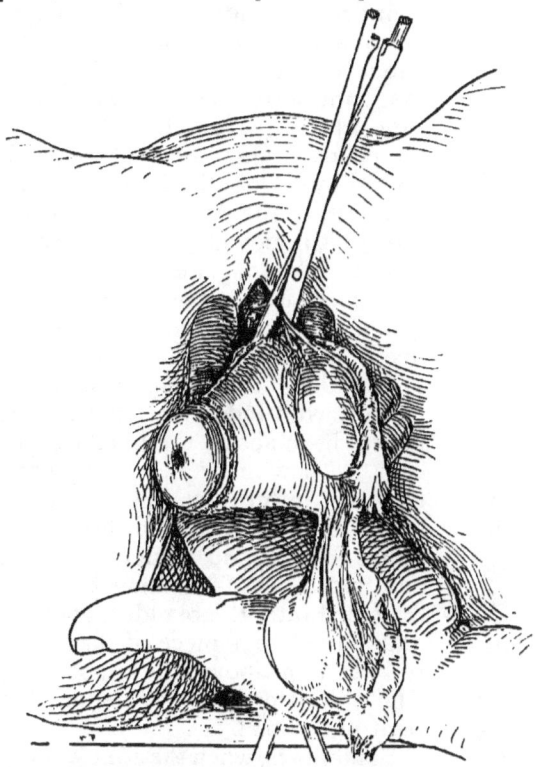

FIG. 70.—Having clasped the right ovarian artery, the uterus is cut away upon that side. The operator rotates the uterus, so that the cervix is delivered and the posterior surface of the uterus presents. He grasps the left broad ligament between his index and middle fingers, and applies the forceps to the left ovarian artery. The method of applying these forceps is shown (photograph of operation).

tion that the bite of each pair of forceps is in the upper part of the vagina. No forceps, if possible, should ever be applied so as to project up into the pelvic cavity among the intestines. The gauze pad supporting the intestines

PELVIC INFLAMMATION.

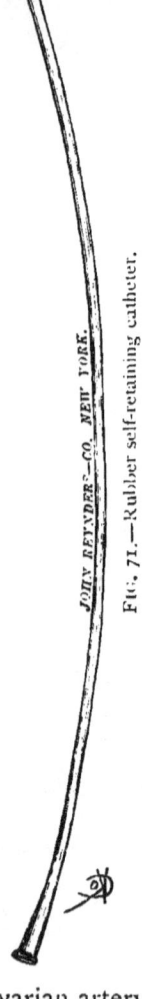

Fig. 71.—Rubber self-retaining catheter.

is now removed, and a piece of iodoform gauze is inserted between the forceps and the wall of the vagina on each side to prevent pressure-slough. The operator now takes squares of iodoform gauze, each about two inches wide and three inches long, and introduces one piece along the side of the forceps on the left, a little above their tips. This piece of gauze is supported by a smooth, narrow speculum introduced to the right of it; the dressing forceps is removed, another piece of gauze introduced alongside the speculum, the speculum withdrawn, and this piece of gauze also supported. In this way the operator proceeds from one side to the other, filling the opening in the vagina entirely with iodoform gauze, which projects a little above the points of the forceps. A few more pieces of gauze are introduced lower down in the vagina, so as to fill it to the vulvar orifice. Sterilized gauze is wrapped around all the forceps and tied. A self-retaining catheter is introduced into the bladder upon a sound and pinned to a piece of plaster fastened to the skin above the pubes (Fig. 71). The sphincter ani is dilated, and the patient put to bed.

Sometimes, when the adnexa of one side are so firmly attached to the intestines, or are so large, or the vulva is so small that the operator cannot loosen both adnexa to his satisfaction, he may proceed as follows: If the difficulty be limited to one side only —for example, the right side—he may free the adnexa on the left side, secure the uterine arteries on both sides, and the ovarian artery on the left side outside the ovary and tube;

he then cuts the uterus free on the left side. Having done this, he introduces a pair of forceps close to the uterus upon the right side where the adnexa have not been freed, and removes the uterus and adnexa of the left side, leaving in the tissues which embarrassed him. It will now be found that he will have room for removing the remaining adnexa under the guidance of the eye. To do this the operator will secure the ovarian artery outside the ovary and tube. This will render the forceps which was applied between the uterus and right adnexa unnecessary, so it may be removed. This is in reality but a form of morcellation or removal in fragments.

While the removal of the uterus *en masse* is more generally accepted than any other method, I am persuaded that it is responsible for many of those ill results which lend arguments to the opponents of the vaginal method. In certain cases it is utterly impossible to remove the uterus and adnexa entire. Such cases are those where the uterus is much enlarged, where the pus foci are enormous or attached high at the pelvic brim, and cases of advanced genital sclerosis. It may be found impossible to free the adnexa before applying forceps, and equally so after forceps have fixed the tissues.

Appreciating the difficulty of ablation *en masse*, I have for several years practised exclusively ablation by hemisection. This I sometimes supplement by morcellation, but the morcellation is employed merely as a step preliminary to the hemisection.

ABLATION BY HEMISECTION.

"I divide my difficulties by splitting the uterus." This is the operation which I always employ. It is the operation of election in all cases, whether associated with fibroid degeneration or not. In such cases it is sometimes associated with, but never supplanted by morcellation. By means of this procedure, the time consumed in operating is rarely twenty minutes, and the operation is always complete. Remembering his anatomy, the operator recalls that both upon its anterior and posterior surfaces, the uterus is comparatively sparsely supplied with vessels, along the middle line. Therefore, an absolutely median section produces but little hemorrhage. The time of operating is short, for, by means of the hemisection, each set of adnexa and its corresponding half of the uterus are rendered movable. Further, as one-half of the severed uterus is shoved up into the pelvis, out of the way, the hand is enabled to work high in the pelvis to the side of that half of the uterus which is drawn down, and the fingers have all the space to one side in which to work, from the bladder to the perineum. One other advantage is that, as each half of the uterus is liberated and drawn down, it is swung outside the vaginal outlet, giving an unobstructed orifice in which to work.

Operation.—*First Stage.*—The patient is on the back and in the lithotomy position. A short Jackson speculum draws down the perineum. The uterus is curetted and swabbed out, but not packed. The intra-uterine traction-forceps is introduced, and the posterior cul-de-sac is opened (Fig. 72). All adhesions posterior to the uterus along the middle line are severed by the examining finger up to the fundus. No attempt is made to further separate the adherent organs at this stage. It can not be now properly done and is a waste of time. The posterior incision is carried around the cervix, almost to the middle line. Drawing down the uterus and holding up the bladder, the anterior cervico-vaginal juncture is severed by means of the scissors. This cut is not to be made close to the external os, but is above the dense cervical structure, and in the loose pericervical tissue. The fold at which the incision is made is easily seen when the uterus is shoved up. This incision is carried laterally toward the posterior cut, but stops one-eighth inch from it on each side. As this incision is made, a few fine arterioles spurt. They are not important, being but small anastomotic branches between the uterine

FIG. 72.—Showing the method of incising the vagina at the point $\times$ in Fig. 62. The intra-uterine traction forceps is shown pulling the uterus down. The second step in all vaginal ablations (from a photograph of an operation by the author).

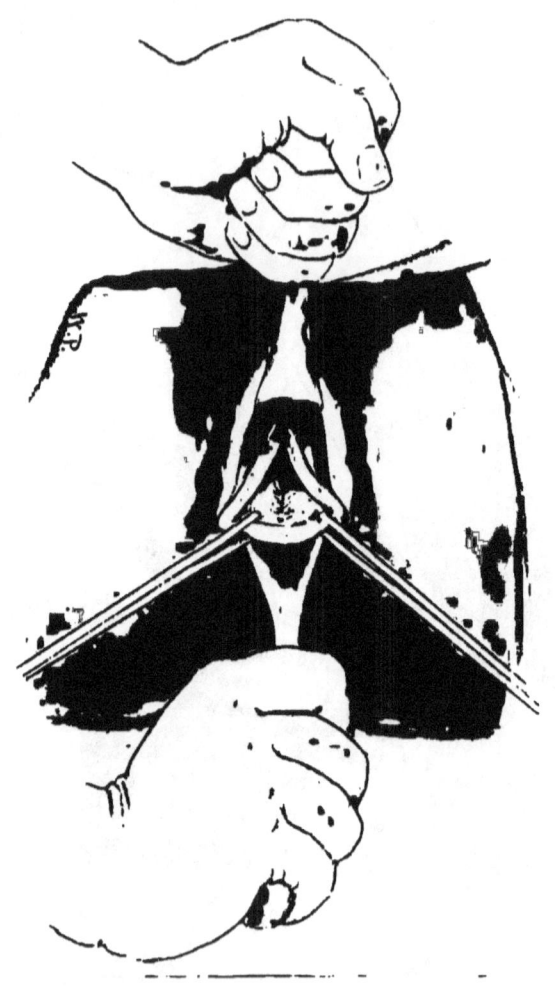

FIG. 73.—The cervix is split anteriorly. The first step in hemisection (from a photograph of an operation by the author).

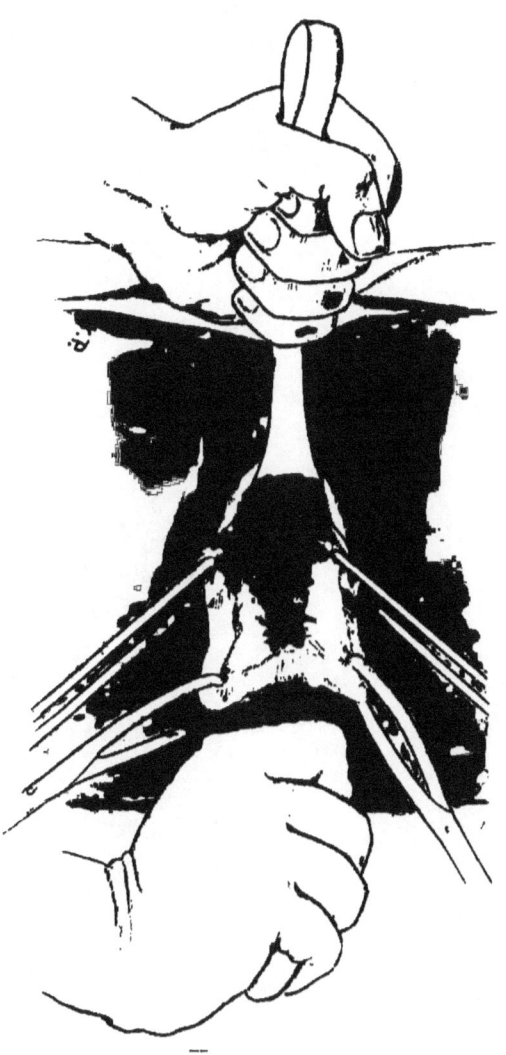

FIG. 74.—Showing the effect of splitting the anterior uterine wall so that the uterus may be rolled from beneath the bladder (from a photograph of an operation by the author).

and vesical arteries. The assistant digs the *short*, narrow Jackson retractor into the anterior cut, with the edge planted hard against the cervix. This is the way to avoid wounding the bladder. The uterus is then rotated by twisting the intra-uterine traction forceps, and the operator is thus enabled accurately to determine the loose tissue between the bladder and uterus which he is to cut. As these fibers are severed with scissors the retractor draws the bladder further and further upward, while the uterus sinks lower as it becomes free. Another simple way of severing the connection between the uterus and bladder is by blunt dissection with the finger. In doing this, the traction forceps is invaluable, as it furnishes a hard surface upon which to press. This dissection is made by shoving up the pericervical tissues with the finger pressed hard against the cervix. The point of the finger is never allowed to wander away from the uterus. If this rule is adhered to, the bladder will not be wounded. The dorsum of the finger is upward, and the actual dissection is effected by a sort of rubbing motion with the palmar face of the end of the finger: the bladder is rubbed off the uterus. If the incision is made close to the external os, this dissection is most difficult; but if made as I suggest, it is easy. After the uterus is free in front and behind, the first stage is finished.

Second Stage.—Two pairs of bullet forceps are made to grasp the angles of the external os, and the intrauterine traction forceps are withdrawn. The two index fingers are introduced between the bladder and the uterus, and the bladder is further separated from the uterus to the sides of the latter. This will remove the ureters from all possibility of injury. The bladder is held up out of the way, while assistants draw down on the bullet forceps. The blunt scissors are inserted as a sound to determine the direction and shape of the uterine cavity, and are then withdrawn. As far up on the anterior surface as the operator can see, he splits the uterus in the middle line. The assistants evert these edges by twisting the bullet forceps outward, and the upper end of each side is grasped with French traction forceps (Fig. 73). As these are drawn upon and rotated outward, it will be found that more of the uterine body comes into view, and is unfolded so that the uterine cavity is flattened out. All of the uterine cavity that can be seen is split in the middle line, and other traction forceps are entered higher up. In this way the fundus is reached and severed (Fig. 74). All specula are now withdrawn,

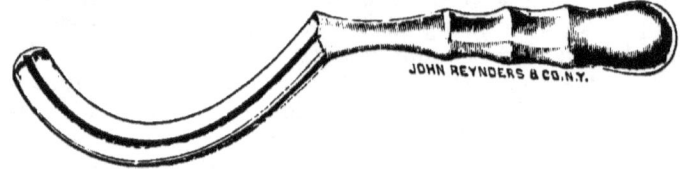

FIG. 75.—The author's retracting grooved director. Of great service with soft, friable uteri.

and my grooved director (Fig. 75) is introduced behind the uterus, entering behind the cervix. A finger is inserted behind the bladder and the director is felt; and again the finger is forced behind the uterus to see that no guts lie between the director and the uterus. The assistant is told to press down the perineum hard with the director,

FIG. 76.—Bistoury for splitting the uterus.

while the curved portion of the instrument pulls forward the uterus. A short speculum is inserted behind the bladder until the groove in the director is seen. Into this a special bistoury is inserted (Fig. 76), and the uterus is split accurately in two halves. This completes the second stage (Fig. 77).

Third Stage.—The director is drawn out. The right (on operator's left) adnexa and half of uterus are shoved into the pelvis, while traction is made upon the left half

FIG. 77.—The anterior wall of the uterus has been split until the organ has been rolled from beneath the bladder. The author's grooved director is shown circling the uterus and the bistomy is in place preparatory to the final step in hemisection.

of the uterus. Did I not leave a narrow strip of vaginal mucosa upon each side when I shove up this half of the uterus, the uterine artery would be torn from its bed and its branches to the cervix broken. After this has been

FIG. 78.—The uterus having been split into halves, one half is rolled out beneath the bladder, and the hand is thus allowed to enter the pelvis. All adherent organs can be liberated even though attached to the pelvic brim. Note the absence of retractors and artery forceps (from a photograph of an operation by the author).

turned out from beneath the bladder, it is swung to the patient's left, and all of the left hand except the thumb is inserted into the pelvis (Fig. 78). The left adnexa are readily liberated from all adhesions behind the broad ligament, as the operator can reach the pelvic brim. If the vagina is relatively small, the operator allows the left half of the uterus and the free adnexa to escape into the pelvis, and draws down the right half of the uterus, and liberates the right adnexa (Fig. 80). But if there is ample room, after freeing the left adnexa, they are drawn in front of the cornu and a forceps is applied from above downward outside the ovary (Fig. 81). *This is the first attempt at hemostasis.* The broad ligament is cut to near the ends of the forceps, and then the uterine artery on that side is clamped from above downward or from below upward close to the cervix, as may be most convenient. The points of the two forceps lap, the one on the uterine artery being exterior to that on the ovarian artery when put on from below, but internal to the ovarian forceps when applied from above (Fig. 79). In this way splitting of the broad ligament is avoided, and when the upper forceps is dropped it will lie alongside the lower. In dropping this forceps (Fig. 82) the upper portion of the broad ligament is folded over the forceps on the uterine artery, and this forceps is kept from touching the bladder. Further, the weight of the upper forceps positively keeps the ovarian artery stump on a level with the uterine, and at the vaginal vault.

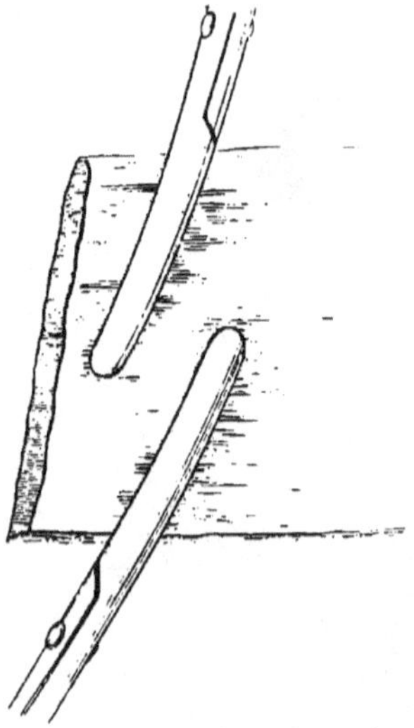

FIG. 79.—The forceps on the uterine artery has been placed from below, while that on the ovarian artery has been placed from above. It will be noticed that the points lap.

The methods pictured in Figs. 83 and 84 are both faulty; the latter for the reason that the forceps will tear the ligament when dropped, and the former because some risk is run in putting the upper forceps on from below

FIG. 80.- The right adnexa are shown drawn out of the pelvis preliminary to applying the first pair of forceps to the right ovarian artery.

FIG. 81.—The right adnexa are drawn across the face of the right half of the uterus and forceps is being applied to the right ovarian artery.

upward, because (*a*) the points of the forceps project too high in the pelvis among the intestines, and (*b*) the ovarian artery is insecurely grasped. The left half of the uterus is cut loose, and removed together with the left adnexa. The adnexa and that half of the uterus upon the right side are similarly treated.

The relation of the ureter to the cervix is greatly modi-

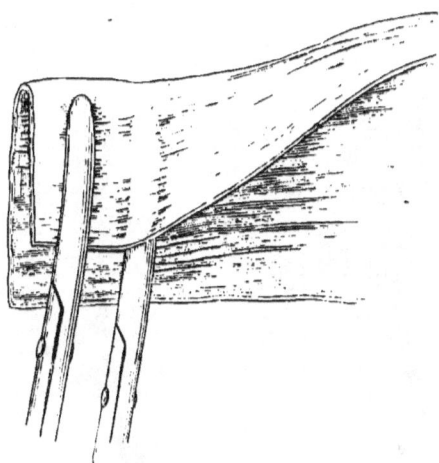

FIG. 82.—After the forceps are applied and dropped down, the upper forceps on the ovarian artery drags down the broad ligament so that it is folded over the lower forceps on the uterine artery.

fied by the hemisection. In applying the forceps to the uterine artery the cervix is sharply drawn to the opposite side. This straightens the curved portion of the uterine artery, and markedly increases the distance between the cervix and the point at which the uterine artery is in relation with the ureter. It will be noticed that no retractors are employed during this stage. They are only in the operator's way. Gauze pads, each secured by a stout string, are introduced into the pelvis above the forceps. The perineum is drawn down by a long Jackson retractor, while the bladder is held up by a trowel. The table is lowered, and a careful inspection is made of the stumps and pelvic contents. If bleeding points are seen, they are grasped; but if the operator has done his work properly, four pairs of forceps are all that will be needed. The gauze pads are removed, and the pelvis is carefully

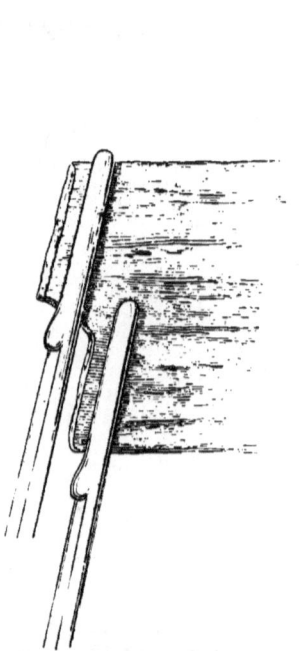

FIG. 83.—Application of clamps from below. Faulty method, as the ovarian forceps projects too high.

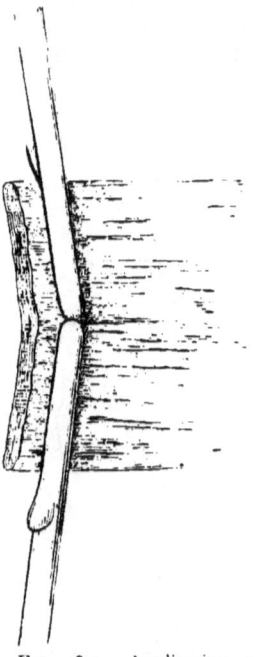

FIG. 84.—Application of clamps from above and from below. Faulty method, as the forceps will not lie loosely when dropped, and will tear the ligament.

cleansed by gauze swabs, particular attention being paid to the cul-de-sac. This completes the third stage.

Fourth Stage.—Dressings.—"The pelvic Mikulicz." A piece of iodoform gauze is inserted between the forceps and the vagina upon each side. Each set of forceps is then drawn toward the lateral pelvic wall by means of a

FIG. 85.—The application of the pelvic Mikulicz dressing. The method of holding the dressing to one side while successive pieces of gauze are introduced is to be noted.

long, narrow retractor. Between them enough strips of gauze are inserted to fill the space. These strips project up above the points of the forceps (Fig. 85). The patient is lowered to the horizontal position, and a self-retaining

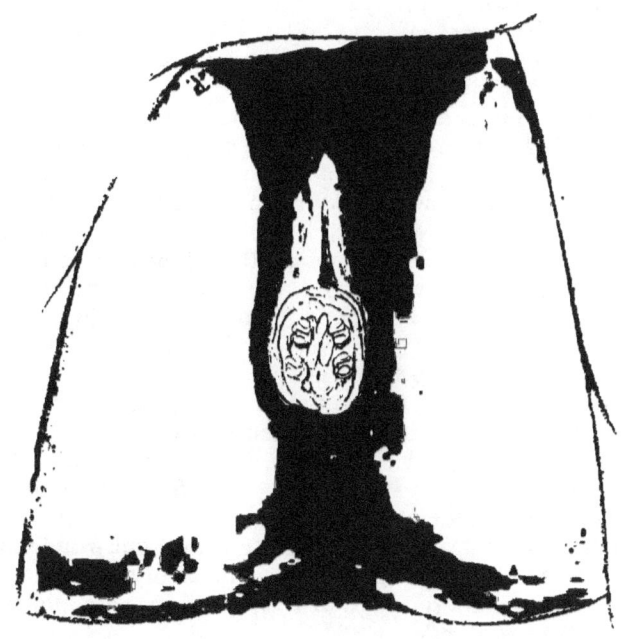

FIG. 86.—The completed operation. The forceps are shown surrounded by the dressings.

catheter is introduced on a sound. The sphincter ani is dilated thoroughly. This is done to allow of the easy escape of intestinal gases, and to allay spasm of the levator ani muscle. The opposing muscle to the levator ani is the sphincter. Under the bruising and stretching to which the levator is subjected, it is apt to spasmodically contract if held down hard by the undilated sphincter. Patients who have the sphincter dilated are more comfortable than are those in whom this is not done. A piece of plain gauze is wrapped around the forceps and tied. The operation is completed (Fig. 86).

The method of making these dressings is radically different from that employed elsewhere. I consider it an essential feature of my method. The Mikulicz dressing is employed here to absorb all discharges. It should be of sufficient volume to do this during the week in which plastic union is taking place between the rectum and bladder. But there is another reason why I pack these cases so snugly. It is to avoid an accident which not infrequently happens to those who use the gauze in slender strips only. When the latter dressing is used, at the time the forceps are removed, the sloughing ovarian stumps very often snap back into the pelvis, causing secondary infection. The pelvic Mikulicz dressing holds these stumps immovably fixed at the vaginal vault, and I have never seen such secondary infection.

In a case of what I supposed was a secondary hemorrhage from an ovarian vessel, when I removed the forceps on the second day, I made a rapid section of the belly. There was even at this early day found firm plastic union between the bladder and rectum, and the field of my vaginal operation was found completely shut out from communication with the general pelvic cavity. The after treatment usual after vaginal hysterectomy is employed.

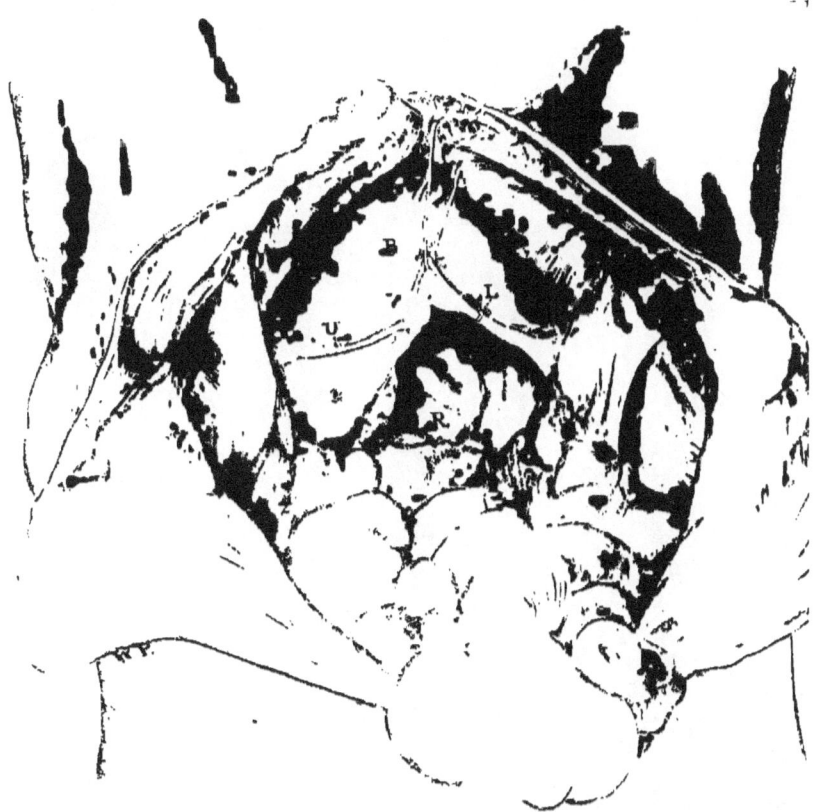

FIG. 87.—Dissection of a body upon whom years before a vaginal hysterectomy had been performed. L, a calcified silk ligature upon the right uterine artery; U, the left uterine artery. No trace of ligature was found on this vessel and the artery still contained a small channel throughout its entire length; B, bladder; R, rectum. The manner in which the vault of the vagina becomes closed by a thin transverse line of union is well shown. Notice how the bases of the broad ligaments hold up the vagina. There is no tendency to hernia, and the posterior cul-de-sac is just as deep as ever it was. This specimen is of value to us as showing the manner in which the vaginal vault continues to be supported even after removal of the uterus.

MORCELLATION.

The uterus is removed in fragments by a process of decentralization.

There are certain cases of very large ovarian abscess which pin the uterus up against the symphysis and immovably fix it there. In such cases the anterior peritoneal space cannot be reached until the uterus is either split or partially cut away as the abscess is evacuated. All broad ligament accumulations demand either hemisection or morcellation. Such are broad ligament abscess, broad ligament hematoma produced by ruptured ectopic gestation, broad ligament cyst of large size, and fibroids with intraligamentous nodules. Morcellation is here necessary because the uterus is displaced so far upwards or to one side and the pelvis so blocked that to even feel the ovarian region the uterus must be removed. In such cases the morcellation is atypical. The peculiar relation of these broad ligament growths to the posterior cul-de-sac must be remembered (see *Exploration*).

The necessity for morcellation is not usually determined until the attempt at removal by hemisection has been found impracticable. Indeed, by whatever method the ablation is attempted, a resort can always be had to morcellation. A most accurate knowledge of the minute and regional anatomy of the parts is needed for this operation. Remembering that the blood supply of the uterus approaches the cervical and cornual points and has lateral anastomoses between the upper and lower vessels, and that the arteries which course across the anterior and posterior surfaces of the uterus are small, the operator feels secure in severing all tissue which lies between the lateral ovarian-uterine anastomoses. The object in doing this is to so weaken the tissue in view that more can be pulled down from above by the process of decentralization, or removing the center, and allow of diminution of the bilateral diameter of the organ. There are two chief ways of doing this. The one most successful in dealing with large uteri associated with pus (the condition we are discussing) is to weaken the anterior uterine wall by removing successive vertical strips of tissue. Mere fixation of the uterus is no indication for morcellation; the fixation must be accompanied by marked enlargement. Typical or symmetrical morcellation is rarely possible when dealing with pus cases, the operator often combining several methods in excavating the uterine wall.

Operation.—It is a great aid if the posterior cul-de-sac can be opened. This is first done; next the bladder is dissected from the uterus until the anterior peritoneal pouch is opened up as far as is possible. While the bladder is held up by a Jackson speculum and the intestines protected by a gauze pad, the anterior wall of the uterus is split as high as possible. Holding the everted edges of the cut with bullet forceps, the operator trims a strip of tissue about a quarter of an inch wide, first from one side, and then from the other (Fig. 88, 1 and 2). A half-inch has now been taken out of the entire visible anterior uterine wall. The removal of this amount of tissue from the cervix will usually be all that can be taken away without reaching its sides. The other slices cut out will be above the cervix and limited to the body of the uterus. In most cases it will be found that the removal of the first two strips has so weakened the anterior uterine wall that the median splitting of the anterior wall can be continued, and the cornua uteri can be brought into view beneath the bladder (3 and 4 of Fig. 88). But in some cases the bladder is attached so high up upon the uterus that the dissecting finger can not effect the separation. Then it will be necessary to split the uterus up as high as possible and remove from each side one, and perhaps two wedge-shaped pieces with their bases upward (5 and 6 of Fig. 88). The stumps are firmly grasped and the anterior wall pulled further down, while the bladder is pushed up so as to expose more of the uterine tissue. What appears is again split in the middle line, and from each side a wedge of tissue is removed (7 and 8 of Fig. 88). Progressively pulling down the uterus and cutting out pieces, the cornua appear. So far there has been free capillary bleeding, but none from vessels of large size. There has been no hemostasis. When the cornua come into

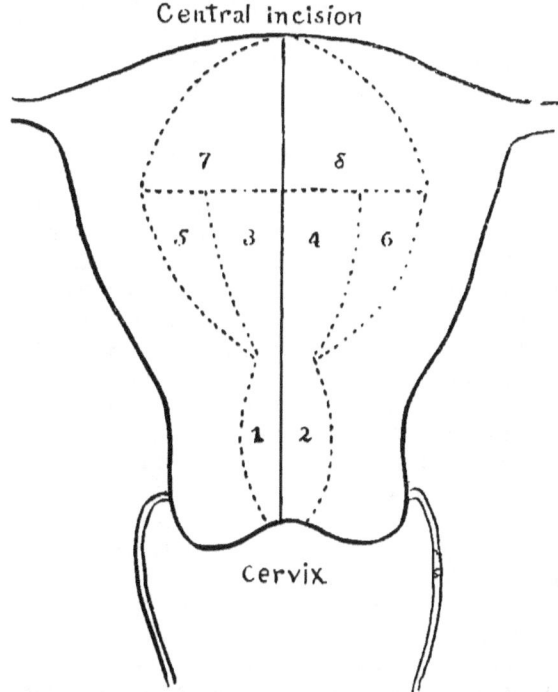

FIG. 88.—A scheme of symmetrical morcellation. The segments are removed as numbered. Sometimes it will be necessary to remove segments 1, 2, 3, 4 only, and this is especially true in pus cases with hypertrophy of the uterus. But in fibroid cases the procedure will have to be pursued so as to embrace most of the tissue included within the dotted lines.

view, if necessary, a large wedge is cut from the fundus, the base of which is at the top of the uterus. This piece will encroach upon the posterior surface of the uterus, and at once upon its removal the cornua with their tubes come still further into view. The grooved guide is now inserted behind the uterus and the organ split in two parts. The further steps of the operation are described under Hemisection, third and fourth stages. In reality morcellation is not a very important factor in the removal of inflamed uteri. In fibroid extirpation it is an invaluable essential. In these pus cases the morcellation is useful only as a step preliminary to hemisection. Without it in certain cases hemisection is difficult. If there be absolute fixity of the cervix, such as we see in bilateral broad ligament abscess, it will be necessary to secure the uterine arteries and cut the cervix loose at the sides before beginning with the hemisection and morcellation; but I consider it a misfortune when I am compelled to apply hemostasis before the adnexa are free (Fig. 89).

Sometimes the operator will find that even after he has removed all the visible portion of the anterior uterine wall he can not turn down the cornua beneath the pericervical ring. Either the adhesions above the uterus are so dense that the cornua are fixed, or else there is a mass behind the lower zone of the uterus which prevents descent of the organ. When he comes to a standstill in his anterior morcellation he proceeds as follows: The uterine arteries are clamped by two forceps and the cervix is freed with scissors. The cervix is then amputated at the level of the internal os. A firm grasp is taken of the stumps, and the posterior uterine wall is morcellated as was the anterior. After proceeding half way up the uterus in this manner it may often be found that the uterus is so shelled out that it may be partially inverted, or that one cornu may be brought so far into view that forceps may be applied to the ovarian artery close to the cornu. If this can be done it is an easy matter to cut the uterus loose upon one side and to swing the mutilated organ out of the vagina. The enucleation of the adnexa attached to this large portion of the uterus is then made as though the uterus had been split in half, and it is removed with the adnexa of that side as in hemisection. Then the adnexa outside the forceps which was first placed on the ovarian artery of the opposite side is freed and brought out; the ovarian artery secured outside the ovary and the adnexa together with

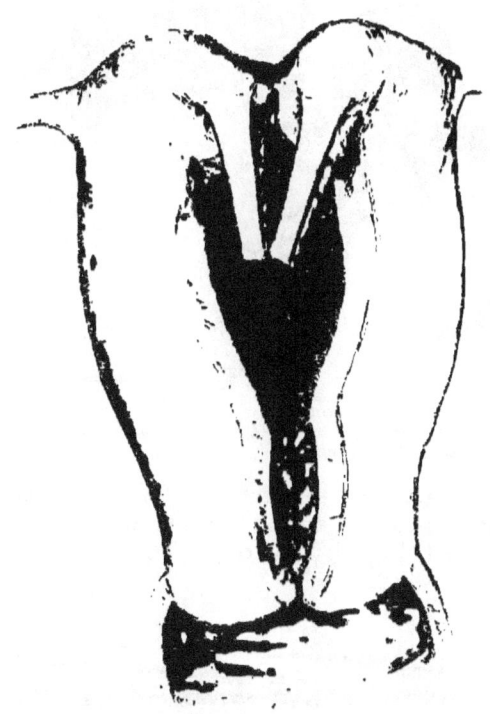

FIG. 89.—Showing the effect of morcellation as outlined in Fig. 88. The traction forceps draw the cornua together so that the fundus is made to roll out beneath the bladder.

the provisional forceps are removed. By this method of irregular morcellation very large uteri can be taken out through the vagina with the use of only four forceps

(Fig. 90). All through such an operation as described repeated palpations must be made of the arterial anastomoses at the sides of the uterus, and the utmost care must be exercised not to wound them either by scissors

FIG. 90.—Symmetrical morcellation of the fibroid uterus. The uterus reached the level of the umbilicus. Miss L., net. 42. Vaginal Ablation. Four pairs forceps used. Recovery.

or traction forceps. I employ for morcellation very stout scissors curved on the flat, one blade blunt and the other pointed (Figs. 91 and 92). The pointed blade can

FIG. 91.—Sharp heavy scissors, especially useful in morcellation. Either point can be driven into the tissues, however hard.

be driven into the tissue. I am further careful always to cut from without in.

An assistant can lend material aid by pressing down above the pubes, employing the closed fist for this purpose. Such support prevents the uterus being drawn up in case the traction forceps tear through. The operator should strive to avoid lacerating the uterine tissue by pulling his traction forceps through it. This may happen to him once, but the one experience should teach him the degree of traction the tissues will tolerate without tearing.

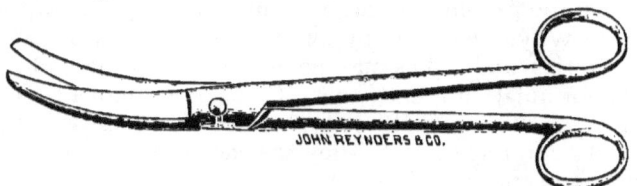

FIG. 92.— Stout blunt scissors used in vaginal hysterectomy.

Like all very technical manœuvres pages of description of the various steps do not become mental pictures until applied. But one operation upon a difficult case will suffice to make clear the necessity for all I have written. If the operator meets fibroid nodules within the uterine walls they are shelled out of their beds. The removal of each of such isolated growths aids in the progress of the operation.

VAGINO-ABDOMINAL HYSTERECTOMY IN THE PUERPERAL STATE.

Indications.—It is supposed that a possible malarial paroxysm has been eliminated by cinchonizing the patient by means of a rectal injection of quinin solution (see formulæ). Faithful trial of intra-uterine irrigations (see septic endometritis) have failed to subdue the symptoms of septicemia, and the operator determines to open the posterior cul-de-sac. This he does after performing curettage. When the cul-de-sac is open the propriety of performing hysterectomy may be settled, but it is impossible before. Upon inspecting the uterus it is found livid and usually studded with isolated flakes of lymph. The curettage has shown the inside of the uterus to be necrotic, and after the cul-de-sac is opened slight pressure with the examining finger will break the uterine wall. The uterus is in a necrotic condition. The fluid evacuated from the cul-de-sac may be muddy serum containing more or less lymph, or scropus may be present in large quantities lying free in the pelvic cavity. Almost any one of the various lesions of the ovaries and tubes may be found. But the indications for ablation are the necrotic or gangrenous condition of the uterine walls, and a septicemia which will not yield to curettage and cul-de-sac evacuation. The presence of pus in the pelvis with a firm uterus does not call for ablation. A sufficiently effective evacuative operation can be made through the cul-de-sac without removing the uterus; and the presence of enormous amounts of recent lymph not only does not call for the ablation, but rather contra-indicates it, if the uterus be firm. If by bacteriological examination of the discharge streptococci have been found, this is but another reason for hastening the operation. I wish to be clearly understood as opposed to this formidable operation for septicemia where there is absence of signs of uterine necrosis. Streptococci may be present and large quantities of pus produced, and yet the cul-de-sac evacuation will suffice to effect a cure. If the uterus is beyond saving, so long as it remains it feeds the lymphatics with septic material. These women die, not from the peritonitis and pus foci, but from septicemia.

Operation.—Rapidity in operating is essential. The uterus is curetted, irrigated, and packed with gauze. The posterior cul-de-sac is opened, and the bladder is partially separated from the cervix by incising the vaginal mucous membrane. Into the posterior incision iodoform gauze is stuffed. While he cleanses his hands, the operator has the patient placed in Trendelenburg's position and prepared for laparotomy. The abdomen is opened from the umbilicus to the pubis. As soon as it is entered the intestines are gently brought above the pelvic brim, and held there by large gauze pads. Stout hysterectomy forceps are made to clamp the broad ligaments outside the ovaries, and the broad ligaments are severed internal to and alongside the forceps. Smaller forceps secure the spouting ovarian arteries at the sides of the uterus. The operator then lifts the uterus up and draws out the plug of gauze in the posterior cul-de-sac. He inserts two fingers into the vaginal opening, so as to hold the uterus in the palm of his hand, and hooks his fingers in the vagina in front of the cervix. Upon these as a guide he strips the bladder from the anterior face of the cervix. This is done by first making an incision through the peritoneum at the uterovesical fold which extends across the face of the uterus, and then by means of the fingers of the right hand literally rubbing the bladder tissues away from the uterus. As the anterior vaginal wall was incised at the time the curettage was done, it is easy to dissect the bladder from the uterus. Two pairs of forceps which are inserted alongside the uterus are made to grasp the central portions of the broad ligaments. The ligaments are cut to near their points. Two other forceps are made to grasp all the remaining tissues at the sides of the cervix. These last forceps secure the uterine arteries, and their points project into the vagina. The uterus is then removed. The pelvis is wiped dry, and the vagina is packed with gauze from above. The six forceps are lifted up in a bunch and are carefully surrounded by iodoform gauze strips, which project above the skin of the abdomen. At no point must the forceps touch the pelvic floor or lateral pelvic walls. They will produce slough wherever they rest. The pelvic dressing is very large, the pelvis being completely filled with gauze. A few stout silver wire sutures are passed so as to approximate the peritoneal, fascial, and muscular planes around the gauze. The abdomen is dressed in such a way that no pressure can be brought upon the forceps. This is most important. Into the elbow vein from thirty to ninety ounces of normal salt solution are injected. If the patient lives, the forceps are removed in forty-eight (48) hours. Under chloroform all dressings are changed in one week, great care being exercised to fill the pelvis. In doing this the intestines are held back by retractors, and as a soiled strip of gauze is removed, a clean one is inserted. The intestines must not be allowed to leak down into the pelvis. After removing the abdominal dressings, the patient is placed in the lithotomy position and the vaginal dressings are renewed. Other dressings are made as indicated. The outpouring of serum at first is enormous. After the first dressing the production of pus is pronounced. It is dirty surgery, but it is life-saving. To use ligatures is to make sinuses even if the ligatures hold in the rotten tissues. To close the belly is to lose the patient. Not alone the uterus, but all the retroperitoneal tissues are infected, and provision must be made for the escape of their septic contents. The gauze is not used alone for drainage, but to isolate the entire pelvis. Iodoform poisoning may occur, but it is a risk which must be encountered. To save even fifty per cent. of these women is a triumph, for all would die without the operation. Those which have died were those in whom the tedious ligature operation was done and the belly closed. It will be noticed that both pelvic and abdominal Mikulicz dressings are made.

The *general treatment* is important. Hypodermics of strychnin, gr. $\frac{1}{50}$, are given q. 3 h. for a day and then diminished gradually. If the *kidneys are damaged*, glonoin, gr. $\frac{1}{100}$, hypodermically, is given q. 3 h. or as often as needed, and the day after the operation another intravenous injection of salt solution is made without narcosis. Having met and cured the most desperate cases of puerperal fever at all stages of the disease, I am warranted in advising this radical work where irrigation, curettage, and cul-de-sac evacuation fail. I cannot sit by and fill a woman's stomach and skin with drugs when I know that she holds within her body a rotten mass filled with myriads of germs each reproducing millions a day. These cases of puerperal fever energetically treated *as soon* as seen, will rarely die. Only those vicious infections, as gangrene of the uterus and thrombophlebitis, will resist the irrigation, or the curettage and cul-de-sac incision. Fortunately such cases are rare, but when met with hysterectomy alone will save them. Too many hundreds of women die in America every year because the let-alone policy is adopted. The higher the authority —speaking against the surgical treatment of this essentially surgical disease—the greater the mischief, for weak brothers applaud high authorities when they preach inactivity. But I desire to utter a warning against the application of this operation in cases which do not strictly demand it. It is in those sudden virulent infections due to streptococci that I have practiced this operation. Such a case will from the first carry a temperature rarely below 103° and a pulse more often above than below 130. And, as I have said above, the curettage and examination of the uterus through the cul-de-sac will demonstrate that the soft, flabby, friable, and discolored uterus is in a condition of cellular disintegration. If even the most virulent infections are treated surgically from the first, it is doubtful if a hysterectomy will be required. It is in the early application of the curettage and cul-de-sac operation in those cases which do not yield to intra-uterine irrigations that we must find a substitute to hysterectomy, or rather a means to render it unnecessary.

AFTER-TREATMENT OF HYSTERECTOMY AND VAGINAL SECTION.

General.—If the patient has been properly prepared for the operation, I give absolutely no drink or nourishment for six hours after the operation. In debilitated women and those profoundly septic, it is advisable to administer fluids at this time. I give one ounce of cold sterile water with five drops of lemon juice every hour. As a rule, this will allay vomiting, and, in those women who have regurgitation of bile into the stomach (green vomit), this is particularly useful. The acidulated water tends to check the vomiting, and seems to cause the bile to flow in the right direction. At any rate, these cases often have several bile-stained stools in twelve hours after the administration of the acidulated water is begun. But my general rule is to keep the stomach perfectly empty for twelve hours, and then begin the administration of either rubinat or apenta water. This is given in half ounce quantities, each dose being followed by a half ounce of sterile water. This is administered every hour, until six ounces are taken. Two hours after the last dose a small enema is given composed of a half ounce of glycerin and five ounces of water. In those cases which have green (bile) vomit I do not give salines until all vomiting ceases. Women with alcoholic stomachs who vomit even in spite of the acidulated water must be given a little iced champagne, about one ounce an hour. The entire object of the first after-treatment is to get the bowels open, and, at the same time, to prepare the stomach for the reception of food. After the bowels have operated, I give a half ounce of hot chicken broth, every hour or so, for the first day, with a bite or two of toast occasionally. The third day I allow coffee and toast in the morning, four ounces of broth and toast at eleven, scraped beef at two, more broth at five. Between the feedings I give abundance of water. Gradually the patient gets upon regular diet, with the exception of fruits and vegetables. These I do not allow until after the first dressing. The third night after the operation, I generally give one compound rhubarb pill, followed next day by a small enema. After the first dressing, I give cooked fruits, meat, soups, potatoes, rice, simple puddings. But all the time, abundance of water. I never give milk. During convalescence cream and oatmeal or hominy are allowed. The prepared foods in the market are useful for rectal injection only, where the stomach refuses to retain anything.

Catheter.—Every two hours after operation, the self-retaining catheter is opened and the quantity of urine escaping is measured. A specimen is analyzed. On the second day the bladder is washed out with saturated solution of boric acid, and the catheter is withdrawn. The urine is drawn every four hours after this. Sometimes the bladder leaks around the stationary catheter, puzzling the inexperienced.

Anodynes.—These I never use except with epileptics. Then I give a little morphin. Cavity work and morphin are incompatible. The pain is pretty severe. It is a new kind of pain to the woman, but is easily endurable. Any relief obtained from the use of morphin is but temporary; it is borrowed. It must be paid back later in vomiting, tympanites, repeated enemata, etc. After the bowels operate the patients are quiet. Usually five hours of refreshing sleep are obtained the second night. The sick-room is to be kept quiet, no visitors being admitted; particularly none of the family, until the patient is out of danger, on the fourth day. In operating in private, if the surroundings are controlled as they are in hospitals, the results will be the same. After operations on women who have the opium habit I give hypodermically morphin gr. $\frac{1}{8}$ and hyoscyamin gr. $\frac{1}{100}$. It is seldom necessary to repeat this.

Position of Patient.—This is generally dorsal, the knees drawn up and supported on a hard pillow. After the forceps are removed, the patient is kept on the back for six hours longer, so as not to cause bleeding by moving, and after this time she is allowed to turn on her side.

Removal of Forceps.—This is done at the end of forty-eight hours. Selecting the lowest forceps, the keys are applied and the forceps unlocked. The catch is separated one-quarter inch. The operator will now appreciate the importance of having his forceps all made alike, for the separation at the lock will tell him the distance between the points which are hidden within the patient's body. The keys are removed and the forceps is twisted about 10° one way, held in that position a moment and then twisted 10° in the opposite direction. Usually this will suffice to loosen the stump from the forceps.

While twisting the forceps back and forth gentle traction is made upon the instrument. No force must be used. If the forceps does not slip out readily, either the stumps are stuck to it and must be freed by repeated twisting of the instrument, or else the gauze has become stuck to the forceps and must be liberated by introducing a blunt, flat instrument of some sort between the gauze and the forceps. In this manner each forceps is removed.

Time in Bed.—Cases of cul-de-sac exploration and replacement are allowed out of bed after the third dressing. This is true also of hysterectomy cases, unless the vagina be widely opened and the perineum gaping. Inasmuch as the third dressing is usually made on the seventeenth day, the patients are out of bed generally before the expiration of three weeks. I make no attempt to hasten their discharge, but allow the surfaces to heal without much physical effort by them, and their minds to recover from the disturbance incident to facing and enduring so serious an operation.

Dressings.—In hysterectomy cases, on the eighth day, I put the patient in Sims' position. While the woman is in this position and perfectly still a careful removal and renewal of the dressings are made. In removing the gauze strips the centre ones are first taken out, so as to loosen those next the vessels. At the top of the cavity will be seen the lymph-covered rectum red and oozing, and upon each side the dead stumps already beginning to blacken. The instruments used in this first dressing are a long-bladed Sims' speculum, my trowel depressor, and a dressing forceps. Sims' tampon screw is a valuable instrument in all these dressings. The second dressing is made a week later.

The method of dressing cul-de-sac and replacement cases is described under the proper chapter.

Behaviour of the Wound.—The sloughs produced by the forcipressure smell badly. There are two ways of removing this: one by douching, after the first dressings are removed; the other, by ample, repeated dressings. I prefer the latter. There is no odor about my patients, although I do not dress them more often than once in a week. The sloughs are blackened shreds and masses at each side of the vaginal incision. They should not be pulled off, but should be allowed to separate gradually. Healing does not really begin until the sloughs have separated, after which it is very rapid. I renew the dressings whenever discharges escape through them, and prefer Sims' position in doing this.

Sometimes in healing there will be produced at the vault of the vagina a knob of granulation tissue. It is better not to make application to this, but to pull it off with Luer's forceps.

Occasionally gonorrheal urethritis is aggravated, and a coincident cystitis induced by the operation and catheterism. Repeated irrigations of the bladder by saturated boric acid solution will correct the latter, and silver nitrate, grs. v. to f$\tilde{3}$i once a day, will cure the former.

ACCIDENTS AND COMPLICATIONS.

Bladder.—The bladder is wounded more often than any other viscus. There are two ways in which the bladder may be injured. In the digital separation of the bladder from the uterus, the finger may enter the bladder cavity. Carelessness in making the separation between the two viscera leads to this. The rent is usually transverse. Upon suspecting such an injury the catheter is passed, and if the mucosa vesicæ is even bruised bloody urine will be withdrawn. Further dissection is effected by the use of mouse-tooth forceps and scissors, as manual violence will but enlarge the opening. Transverse rents in the bladder do not require suture, as they close if the bladder is kept empty.

In that method of separating the bladder from the uterus which is accomplished by progressively dividing the anterior uterine wall and dissecting away the bladder in stages, a vertical rent may be made by the scissors. The contracting bladder tends to keep such an opening permanent. Here, therefore, continuous sutures of fine chromicised catgut should be employed to close the rent and the bladder should be kept empty. The after-treatment is modified by this accident in but one particular, namely, that the bladder should not be allowed to distend. The catheter is left in place a week and is opened every hour. Each day the bladder is irrigated with a saturated solution of boric acid, but at one time no more than an ounce of the solution is to be injected. After the stationary catheter is removed, catheterism is done every two hours and the intervals progressively lengthened. The nurse should be instructed to notify the attending surgeon if no urine flows, for clots may block the catheter. The injection of a little boric acid solution will clear the tube.

The rubber catheter may be pressed so snugly against the pubes that flow of urine through it will be stopped. This condition will be differentiated from stoppage due to clot by rotating the catheter and pushing it up a half inch. If it be pressure obstruction the urine will then flow.

Bowel Wounds.—I have never wounded the gut. But in several cases I have found pus tubes opening into the rectum. After the operation of ablation is completed a continuous suture of fine chromic catgut closes the bowel opening. These openings also tend to close spontaneously, and even an awkward method of suturing will be effective. The rectum should be rendered incontinent in these cases by paralyzing or dividing the sphincter ani.

If the pus sac opens into a coil of small gut, or this latter be wounded, the rules governing the treatment of this accident during laparotomy will apply here. If resection is to be done the attempt should be made to use Murphy's button, but the general rule is that laparotomy and careful suturing are needed to properly close wounds in the small gut.

Wounds of the Ureter.—These are not recognized when made. In fact, ureteral fistulæ commonly occur late as the result of sloughing produced by improperly protected forceps. As the lower half of the pelvic ureter is nourished by vesical arteries, when slough occurs it commonly involves at least an inch of the ureter if produced by grasping the ureter in the forceps; and no method of anastomosis can be applied later on. If such an accident is detected during the operation, laparotomy should be done at once and the severed ends of the ureter be either sutured or the ureter implanted into the bladder. If the ureteral fistula occurs during convalescence, the case should be let alone until no lesion remains other than an uretervaginal fistula. Then the case should be treated as though the accident resulted from laparotomy. At first, attempts are made to close the fistula through the vagina. These usually fail and the surgeon must resort to implantation into the bladder or to nephrectomy. This accident has not befallen me.

Pneumonia.—This occurs not infrequently, as we often operate upon those with phthisis or influenza. The pneumonia commonly develops on the second day, and is of the lobular type. I look with suspicion upon every rise in temperature on the third day and carefully examine the lungs. For this pneumonitis catarrhalis—usually due to streptococcus—I give potassium iodid only, 5 grains q. 1 h. to three doses; stop three hours, and then 5 grains q. 1 h., three doses as before. Until resolution becomes complete, I give 10 grains a day. I have not seen a fatal result from pneumonia following vaginal ablation. The general treatment embraces strychnin hypodermatically and other heart stimulants as needed. So soon as possible the posture of the patient must be changed from the dorsal to the lateral, to check the tendency to hypostatic congestion.

Nephritis.—The method of preparing the patient very much lessens the liability to this complication. For three days after operating all urine is measured and each day an analysis is made. Upon the appearance of symptoms of nephritis, I at once give a high saline enema of three pints. If this is retained, I shall expect to repeat it in eight hours. I also order large draughts of Buffalo Lithia water in hourly administrations. In aggravated cases I give glonoin hypodermatically and use either subcutaneous or intravenous injections of normal salt solution. Digitalis is indicated in cases properly demanding it and should be given as infusion by the rectum. But it is so slow in its action that the diluent normal salt solution must be employed first.

Intestinal Paralysis.—From the first the vomiting is severe and frequently gets worse. Blood may be vomited. The bowels can not be moved and tympanitis becomes marked. The temperature rises, and the pulse becomes quick and weak. The patient is pale, anxious, and in great distress. Neither she nor the physician can detect intestinal peristalsis. This condition I have seen three times, and only in women who had the most firm and extensive adhesion of the *small* gut to the uterus and adnexa, requiring careful dissection to remove them.

The stomach should be kept absolutely empty. Once every three hours one pint of normal salt solution at a temperature of 100° should be injected into the descending colon through a Wales tube (Fig. 110).

Hypodermatically strychnin is indicated, gr. $\frac{1}{30}$ q. 6 h. The first fluid administered by mouth should be a little chicken broth, but should not be given until the stomach has been at rest for twelve hours. In a severe case lasting four and a half days, no food was given, but the patient was kept alive by the salt solution enemata alternating with nutrient enemata. The old treatment of attacking the stomach with cerium, cocain, belladonna, calomel, etc., is irrational. The stomach is normal, and the vomiting is due to intestinal paralysis, regurgitation of bile into the stomach, and reversed peristalsis. Morphia but aggravates the trouble.

Convulsions.—Epileptics and hystero-epileptics will have repeated convulsive attacks. These are best controlled by very minute quantities of morphin, gr. $\frac{1}{12}$ occasionally. These are the only cases in which I employ morphin, and it is indicated because the seizures are due to the traumatism inflicted upon the sympathetic ganglia of the pelvis.

SECONDARY HEMORRHAGE.

Whenever large vessels in the body are secured, either in continuity of tissue or *en masse*, this accident may follow; and the vaginal operation is no exception to this rule. The vessels may be perfectly secure under the forceps, and yet secondary bleeding occur any hour between the time they are removed and two weeks later. The bleeding usually springs from one uterine artery, and is readily controlled by bilateral pressure. (Fig. 93.) A narrow

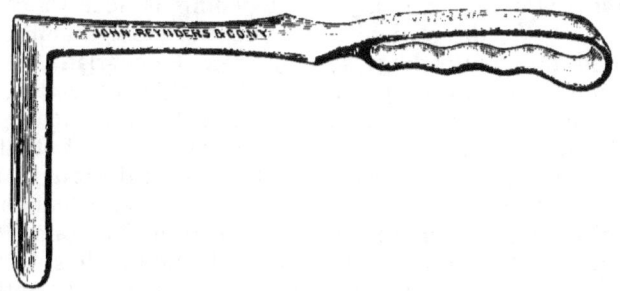

FIG. 93.—Péan's long retractor. I use two in making the pelvic Mikulicz packing.

retractor is introduced through the center of the column of gauze, and one-half of the gauze—that upon the side from which the bleeding comes—is pulled hard against the lateral pelvic wall. A similar retractor is entered alongside the first, and the other half of the gauze pulled to one side. When it is seen that the pressure is sufficient to stop the bleeding, the vaginal packing is increased by the introduction of additional pieces of gauze between the two retractors. The retractor which holds back the gauze over the bleeding vessel is not to be moved until the dressing is complete, but the adjustment and compression of all fresh pieces of gauze are effected by means of the opposite blade. After waiting a few minutes to see whether the bleeding is stopped, the patient is put to bed, the foot of the bed being elevated.

If the pressure does not control the hemorrhage, the patient is placed in Sims' position and given chloroform. All dressings are removed, and the bleeding vessel sought for. Descent of the intestines is prevented by gauze pads, and the bladder is sharply retracted with the trowel, while the perineum is held back by a Sims' speculum. When the spouting vessel is seen, it is grasped with bullet-forceps, which take a firm hold on the tissues, and the stump is lifted away from the vaginal wall. It is then an easy matter to grasp the stump with forceps. The vagina is to be packed with iodoform gauze. If after searching carefully the bleeding is seen to come from above the vaginal vault, and the vessel cannot be found, the hemorrhage springs from an ovarian artery. When the operator is convinced that this is the case, he does not attempt to secure the vessel through the vagina with forceps, nor to compress it with gauze, but, after packing the vagina with gauze to prevent descent of the intestines, he throws the patient into Trendelenburg's position and opens the belly. When he has found the source of the bleeding, the artery is tied with silk and the stump trimmed. The same is done with the other ovarian artery. The ligatures are cut short, and the pelvis cleared of clots. The abdomen is closed. It is well to give a high enema of three pints of salt solution before the patient leaves the table, or to inject sterile filtered normal salt solution into a median cephalic vein. I have seen this accident but once, in one of my earliest cases.

If it be found that the bleeding comes from the azygos artery or other vaginal branch, it is best secured by passing a curved needle around it and tying *en masse* with silk. I can not conceive it possible that so tortuous and long a vessel as the ovarian artery can bleed after its current has been completely shut off for two days. It is probable that the ovarian artery bleeds because the occlusion has been partial and incomplete, and after the removal of the forceps the blood stream bursts through whatever clot has formed in the vessel. It is not so with the uterine artery. After this vessel is clamped, but little of its length remains between the forceps and the internal iliac artery, and, consequently, when the forceps are removed the end of the artery feels the full force of the pressure from the iliac.

I can not explain the very late hemorrhage occasionally occurring when the patient is ready to get up, except upon the hypothesis that the repatency of the artery becomes established. That this does occur I have shown. It has been observed after abdominal hysterectomy with ligature, and has heretofore been ascribed to bleeding from anastomotic vessels. It is always from the uterine artery or its branches, and is easily checked by forceps applied through the vagina.

INTRAVENOUS INJECTION OF NORMAL SALT SOLUTION.

A seven-tenths of one per cent. solution of chemically pure sodium chlorid in soft water is made. This is filtered into either a Florence flask—to be found in all drug stores—or else into a perfectly clean agate kettle. It is then boiled ten minutes and is cooled by placing on ice. The solution is employed at a temperature of 105° F. The infusion apparatus is composed of a twelve-ounce glass funnel, eight feet of pure gum rubber tubing to fit this, and a canula (Fig. 94). The apparatus is boiled twenty minutes in plain water. The hand grasps the arm above the elbow and compresses the veins. The median basilic vein will show running across the bend of

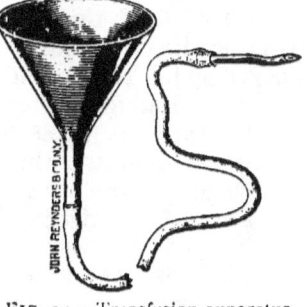
FIG. 94.—Transfusion apparatus.

the elbow from without in (Fig. 95). The skin is drawn upward and is incised carefully alongside the upper border of the vein. Upon rolling the skin down into position the cut is found to be over the vein. The vein is carefully dissected out of its bed. The distal or outer end of the vein is grasped across with an artery forceps, and a half inch internal to this the vein is caught with mouse-tooth forceps. While this is being done an assistant whose hands are absolutely clean, has filled the infusion funnel. This he holds six feet above the patient. The clothing in the patient's axilla has been loosened. The operator severs the vein entirely across and takes the canula in his right hand while holding the bleeding end of the vein with toothed forceps. The saline solution is allowed to flow against the cut end of the vein until the solution feels warm, then the canula is inserted well into the vessel; at the same time, the pressure on the arm is loosed. The assistant watches the flow of water from the funnel, and warns the operator when he is to refill it, so that the operator may compress the tube and prevent entrance of air. To avoid this, all the water is not allowed to flow from the funnel before refilling. The speed of flow is about six ounces in three minutes, or about a quart in a quarter hour. Having introduced the desired amount of fluid, the canula is withdrawn and pressure made around the arm. The two ends of the vessel are secured by fine catgut, and the wound stitched by the same material. Iodoform gauze dressing.

Subcutaneous Injection.—The material is prepared as before. Opposite the angle of the scapula and over the margin of the latissimus dorsi muscle, the skin is cleansed. A few drops of cocain solution is injected, or the skin is frozen with a stick of ice dipped in salt and applied. It is incised for a quarter inch. While the edges are held apart, the solution is allowed to flow through the canula until warm, and the canula is plunged into the cellular tissue between the skin and muscle. Ten ounces of fluid are allowed to enter, when the canula is withdrawn and a stitch of catgut used to unite the surfaces. Iodo-

FIG 95.—The superficial veins at the bend of the elbow (after Quain): 6. The median basilic vein, into which intravenous salt infusion is made. 4. Cephalic vein. 3. Basilic vein 2. Venæ comites of the brachial artery ×. As these latter lie beneath the deep fascia of the arm, they are not in danger in the operation of intravenous infusion of salt solution. (The reader's attention is called to the fact that the elbow vein into which the infusion is made is sometimes the median cephalic, as the veins of the elbow are not constant in their arrangement).

II.

form gauze dressing. Upon the other side a similar injection is made. As the fluid enters the cellular tissue a large swelling appears which subsides in a few minutes. The injection may be repeated lower down in eight hours. I have made three such injections in twenty-four hours in a desperate case of sepsis, altogether sixty ounces. If the fluid is sterile and careful cleansing of skin and apparatus has been made, there is no danger of suppuration following.

The author has observed the following immediate effects of intravenous infusion: the temperature rapidly falls if it has been high, and the pulse has been seen to come from 160 to 110 even during operation. In other words, it is a positive remedy for shock. Remotely, the amount of urine is greatly increased, the specific gravity falls, owing to the dilution, but the actual amount of urea excreted is increased, and albumen, if present, is either markedly diminished or disappears altogether. The procedure is thus particularly applicable to cases of septicemia and hemorrhage. After operation it is demanded whenever the kidneys exhibit evidences of suppression.

INSTRUMENTS.

Short Retractors.—There should be two of these—one narrow, and one broad. I like Jackson's pattern. Two Sims Specula (Fig. 96).

Long Retractors.—One long Jackson and one Péan anterior retractor; two long narrow Péan blades for making lateral pressure; one Pryor Péan trowel (Figs. 97, 98, 99, 100, 101).

Traction Instruments.—One Pryor's intra-uterine traction forceps to be used during the first stage of hysterectomy. Two bullet forceps, strong and with short blunt points. Four French traction, the instruments of Péan. These have four teeth, and behind these upon

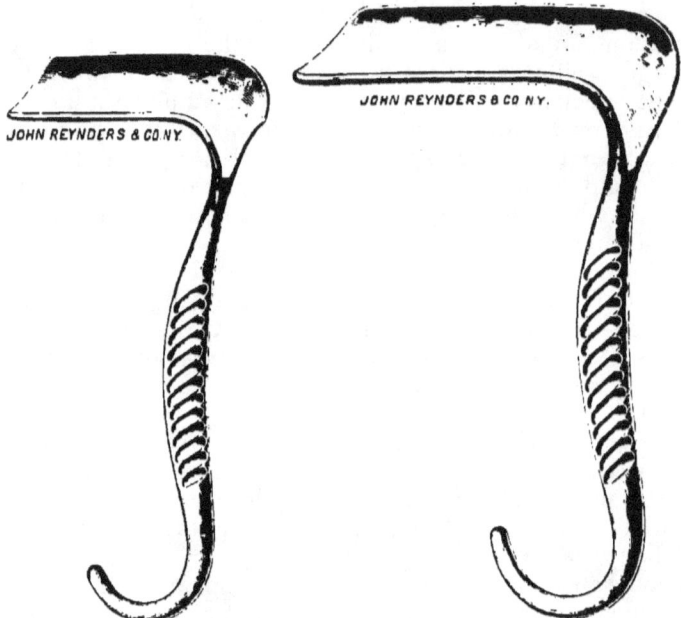

FIG. 96.—Jackson speculum. FIG. 97.—The wide, long Jackson posterior retractor.

each blade are deep serrations. It is important that the teeth look outward when the instruments are open, so as to grasp flat, hard surfaces, as fibroid nodules. The

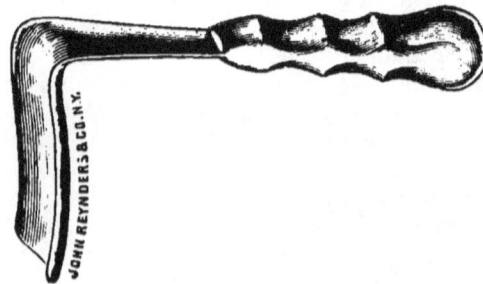

FIG. 98.—Péan's anterior retractor. Useful also as a perineal retractor in small vulvæ.

serrations enable the forceps to hold in soft tissue through which the teeth would otherwise tear (Fig. 102). The short fenestrated forceps of Péan assist in holding the adnexa, and for this purpose Luer's forceps are also valuable (Fig. 103).

Cutting Instruments.—One straight blunt bistoury for bisecting the cervix in stenosis (Fig. 104). Two scalpels with good bellies. One Pryor's bistoury hollow-ground, so as to be easily sharpened. Four pairs of scissors, one Sims vesicovaginal scissors (Fig. 105) to trim ovaries and tubes; one blunt curved on the flat and short; one long, blunt-pointed; and one long

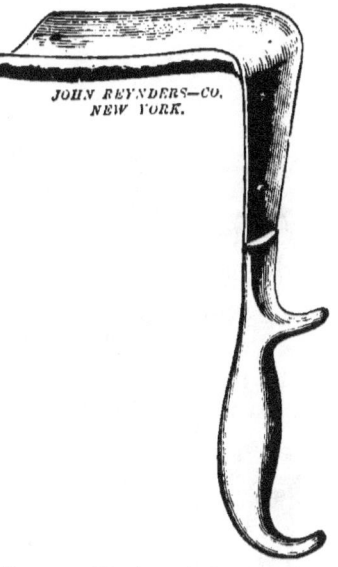

FIG. 99.—Péan's posterior retractor.

FIG. 100.—The author's narrow trowel.

sharp-pointed. Both of the latter have blades curved on the flat.

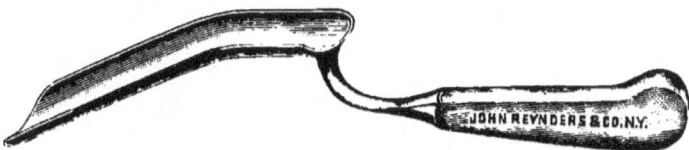

FIG. 101.—The author's wide trowel. Used with women who have large vulvæ.

The **curettes** are the pattern of Sims, and are of three sizes.

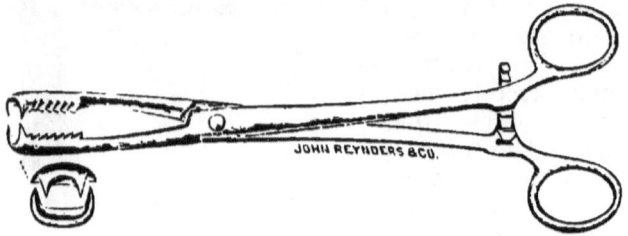

FIG. 102.—French traction forceps. Without them hemisection and morcellation would be most difficult.

The **intra-uterine catheters** are of the Fritsch-Bozeman pattern, and of four sizes.

FIG. 103.—Luer's polypus forceps. The best for holding the adnexa.

As **dressing instruments** I use Sims' tampon screw, a long, slender packing forceps, Hunter's sponge-holder (Fig. 106), and Pryor's packing applicator. Two stout

FIG. 104.—Bistoury for splitting the cervix.

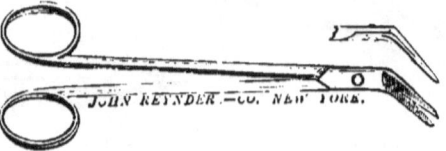

FIG. 105.—Fine scissors for conservative work on the adnexa.

INSTRUMENTS.   235

pairs of mouse-tooth forceps are needed—one long and one short.

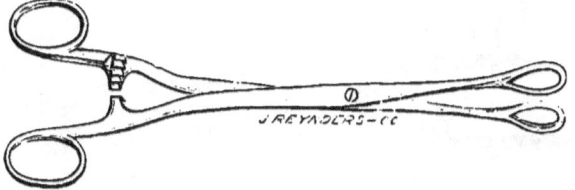

FIG. 106.—Hunter's forceps. I use these for applying dressings.

Hemostatic Forceps.—Four Sims' artery forceps (Fig. 107), eight pairs of Pryor's hysterectomy forceps

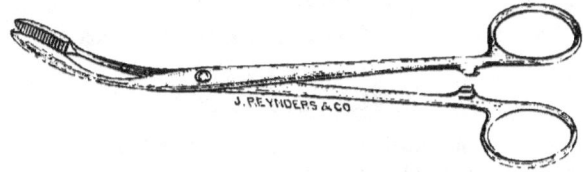

FIG. 107.—Sims's stout artery forceps.

(Fig. 108). These latter are strong, and have transverse serrations. The blades are one and a half inches long.

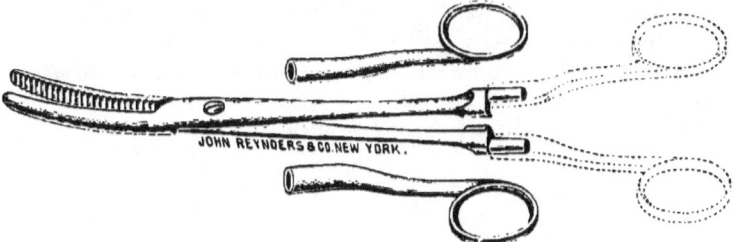

FIG. 108.—The author's forceps. No handles are in the way of the patient's movement.

I use the needle-holder of Sims, and any stout half-curved needles (Fig. 109) with bayonet points. The best instrument for giving high enemata after operation is a Wales bougie (Fig. 110). To secure the patient's legs I employ Ott's crutch.

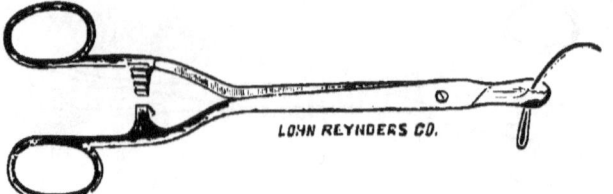

FIG. 109.—Sims's needle-holder. The simplest and best.

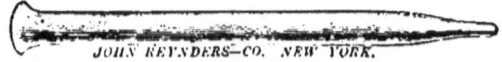

FIG. 110.—Wales bougie for giving high saline injections.

OPERATING TABLE.

I have devised and for a long time used the one shown. It can be employed for any kind of gynecological work, and is portable. (See Figs. 56, 57, 58.) Without it, much of my pelvic work would be most difficult and tedious. The distances in America are so great that those of us who operate over the entire country must go prepared for any work. My table is strong enough for the heaviest woman, and weighs seventy pounds. It is dressed for the operation with blankets and a rubber sheet, or piece of oil cloth. The shoulder brace can slide on the table, so as to support any size of body.

FORMULAE.

Thiersch Solution.—Boric acid crystals, 12 parts; salicylic acid crystals, 2 parts; water, 1000 parts. Tablets to make one quart are sold by Reeder Brothers, Thirty-first Street and Fourth Avenue, New York.

Normal Salt Solution.—This is a .7 per cent. of sodium chlorid in water.

Solution of Quinin.—Quin. sulph. grs. xx, acid tartaric grs. xvii, aquæ ℨiii. Give warm by rectum.

Lysol.—This is five times as antiseptic as carbolic acid, and but one-eighth as poisonous. I use it for my hands as a 2 per cent. solution, and on the patient in a 1 per cent. solution. It soponifies fat and cleanses mechanically as well as chemically.

Iodoform Gauze.—The gauze is sterilized. It is then dipped in a 5 per cent. or 10 per cent. solution of iodoform, in ether, and laid on a sterile sheet to dry. When dry, it is blue and unfit for use. It is now dipped in a hot bichlorid of mercury solution, 1 : 4,000, when the yellow color returns. It is wrung as dry as possible and packed in glass jars. The mouths of these are stuffed with cotton, after which the jars are inverted in a steam sterilizer and sterilized for an hour. The dressing is expensive, but as it is non-poisonous, and requires renewal once where other dressings are changed three times, it is worth the difference. It can be readily made by a careful nurse or assistant. I am using the 2 per cent. and 5 per cent. strengths more than formerly, and find them as good as the stronger.

Chicken Broth.—A fowl is cleaned and skinned. It is chopped into pieces, bones and flesh. These are put into three quarts of water, and actively boiled for eight hours. As the water evaporates, the quantity is kept to three pints by adding boiling water. Strain into a clean bowl and put on ice. This jelly is heated when needed. The flesh of fowl is the only flesh that dissolves in water. This is the first food my patients get after operation.

Beef Juice.—The steak is broiled medium, chopped up and squeezed in a press or lemon squeezer. It is served warm.

Nutrient Enema.—One raw egg, two ounces squeezed beef juice, two ounces milk, one tube of Fairchild's peptonizing powder, warm to 100°. Give this once in four hours.

STERILIZATION.

The Surgeon.—It is exceedingly important that the operator's hands be technically clean, even in dealing with pus cases; but it is difficult to obtain absolutely aseptic hands. The finger nails should be short. The sleeves are rolled up to the biceps, and the hands and arms are scrubbed with hot water and soap. At least five minutes should be devoted to this. This will soften the nail filth. A sterilized sharp steel nail cleaner is used to cleanse the nails. Particular attention should be paid to the base of the nails, as here the loose epithelium is most often found. Not only is all dirt under and behind the nails removed, but the nails should be scraped as well. They are again scrubbed with the brush. The operator hollows his left hand and fills it with chlorid of lime, "bleaching powder." He adds to this a little water and makes a paste in his hand. Selecting a stick of carbonate of soda, "washing soda," he rubs this into the lime paste, and over his hands and arms. The soda is used much as a cake of soap would be. As he continues the process, he will notice that the grains of lime gradually disappear, and when no more grains are present, he puts aside the soda, and washes off the white paste. By this procedure he develops upon his hands and arms nascent chlorin gas, a most powerful disinfectant. Both the essentials can be procured at small cost anywhere. After doing this, the hands are almost certainly clean; but I go further, and scrub the nails and hands in 2 per cent. lysol solution, after which they are rinsed in Thiersch solution. The operator now puts on a sterile gown, and is prepared to operate. While thus preparing himself, one of his assistants who has previously sterilized his hands, has been cleaning the patient's buttocks and vagina. (Vide " Preparation of Patient.") During the operation the surgeon frequently washes in Thiersch solution.

Instruments.—These are boiled in 5 per cent. carbonate of soda solution for fifteen minutes, the knives and scissors being given half this time. The boiling water is poured off, and the instruments allowed to cool or are cooled by cold boiled water. No instrument pans are used, but the instruments are laid out upon a sterile sheet and kept covered from dust. The boiling soda solution not only sterilizes them, but dissolves all fat, pus, and blood upon them. Chemical sterilization, as by formaldehyd gas, is uncertain. I usually boil the nail scrubs with the instruments.

Rubber Goods.—The rubber irrigator (fountain syringe) is half-filled with water and the clip loosened. It, together with the vaginal brush and self-retaining catheter, are boiled in plain water fifteen minutes.

Irrigating Fluids.—I use boiled normal salt solution or boiled boric acid solution 4 per cent. But I have about abandoned irrigation except to wash out large uteri after curettage.

Transfusion Fluid.—Ordinary table-salt is dissolved in soft water to make a $\frac{1}{10}$ of 1 per cent. solution. It is then filtered and boiled in a new kettle, the neck of which is plugged with cotton. This solution is cooled to about 105° F. In handling it, care should be taken not to agitate the contents of the kettle, lest sediment be put in suspension. Whenever there is sediment in the solution, it should be carefully strained through several thicknesses of sterile plain gauze into the transfusion funnel. The gauze may be tied over the spout of the kettle. This is an impromptu apparatus. In my practice I use chemically pure sodium chlorid. The solution is made and filtered into a glass bottle. In this it is boiled. The transfusion apparatus is boiled for twenty minutes in plain water.

Silk, Silver Wire, and Silkworm Gut.—This is rendered sterile by boiling for seven minutes in 5 per cent. carbolic solution.

Catgut and Kangaroo Tendon cannot be prepared by the surgeon as reliably as by several manufacturers. That made by Van Horn, Forty-first Street and Fourth Avenue, is recommended.

Gowns, Gauze, Sheets, Towels.—These are subjected to a continuous column of live steam in a closed chamber for at least one hour. There are several excellent sterilizers, notably the Arnold, sold for a few dollars. If a steam sterilizer can not be secured, the fabrics may be boiled in plain water for a half hour. Before using they should be wrung dry. The gauze and gauze pads may be fastened in bundles in towels, boiled and then dried in a not too hot oven.

Transfusion solution and irrigation are so seldom required, that the preparation for an operation becomes very simple. Every physician should own a steam sterilizer for the preparation of his dressings. A very good one can be procured for ten dollars, just as effective as one costing hundreds.

Hand-basins.—These I always boil in plain water. Perfect cleanliness can not be secured by using basins which have not been sterilized.

INDEX.

A

Abdominal dressings in pelvic inflammation, 61.
Ablation of uterus by hemisection, 187.
    dressings in, 202.
    first stage, 188.
    fourth stage, 202.
    second stage, 193.
    third stage, 194.
    en masse, 180.
    vaginal, 163. See also *Vaginal ablation*.
Abortion, infection after, 34.
    pelvic peritonitis from, 96.
Abscess of broad ligament, 121.
    symptoms, 122.
    treatment, 123.
    ovarian, 108.
    symptoms, 110.
    treatment, 113.
Accidents in hysterectomy, 166.
Acute gonorrheal endocervicitis, 25.
    endometritis, 47. See *Endometritis*.
    salpingitis, 63. See *Salpingitis*.
    peri-ovaritis, 106.
    treatment, 110.
    salpingo-oöphoritis, conservative treatment of, 152.
    septic endometritis, 26. See *Endometritis*.
    septic salpingitis, 69. See also *Salpingitis*.

Adherent ovaries, conservative treatment of, 159.
    retropositions, 117.
    operation for, 118.
Adhesions, separation of, in vaginal ablation, 164.
After-treatment of hysterectomy, 219.
Analgesia in pelvic peritonitis, 99.
Anesthetic, 125.
Anodynes in after-treatment of hysterectomy, 221.
Apoplexy, ovarian, 107.
    conservative treatment of, 159.
    treatment, 112.
Appendicitis, diagnosis of pelvic inflammation from, 57.

B

Beef juice, 238.
Bladder, wounding of, in vaginal ablation, 223.
Bowel, wounds of, in vaginal ablation, 224.
Broad-ligament abscess, 121.
    symptoms, 122.
    treatment, 123.
    cyst, 114.
    conservative treatment of, 160.
    symptoms, 115.
    treatment, 116.

C

Catgut, sterilization of, 240.
Cervical polypi, 22.
    treatment, 24.

Cervix uteri, cystic degeneration
of, 21.
   treatment, 24.
   polypus of, treatment, 24.
Chicken broth, 237.
Chills in pelvic peritonitis, 101.
Chronic gonorrheal endocervicitis, 25. See *Endocervicitis*.
   endometritis, 50. See *Endometritis*.
   salpingitis, 72. See also *Salpingitis*.
   salpingo-oöphoritis, conservative operation for, 155.
   septic salpingitis, 72. See also *Salpingitis*.
Colon bacillus as a cause of pelvic peritonitis, 91.
Conservative treatment, 146.
   of acute salpingo-oöphoritis, 152.
   of adherent ovaries, 159.
   of broad-ligament cysts, 160.
   of chronic salpingo-oöphoritis, 155.
   of cystic ovaries, 158.
   of hydrosalpinx, 157.
   of occluded tubes, 159.
   of ovarian apoplexy, 159.
Constipation a cause of pelvic peritonitis, 96.
Convulsions after vaginal ablation, 226.
Cul-de-sac operation in puerperal endometritis, 44.
Curettage, 126.
   in puerperal endometritis, 43.
   instruments for, 134.
   packing after, 129.
   repeated irrigations in, 133.

Currettage, time for, 130.
Curette, Sims', 135.
Curettes, 234.
Cyst of broad ligament, 114.
   conservative treatment, 160.
   symptoms, 115.
   treatment, 116.
Cystic degeneration of cervix uteri, 21.
   treatment, 24.
   of ovaries, 107.
   ovaries, 107.
   conservative treatment of, 158.
   treatment, 111.
Cystitis, diagnosis of pelvic inflammation from, 58.

D

Diet in pelvic inflammation, 62.
Diffuse pelvic suppuration, 123. See also *Pelvic suppuration*.
Digestive symptoms in pelvic peritonitis, 101.
Douches in pelvic inflammation, 61.
Drainage in vaginal ablation of uterus, 165.
Dressing instruments, 234.
   Mikulicz pelvic, 202.
Dressings in after treatment of hysterectomy, 222.
   sterilization of, 240.

E

Edematous ovaritis, 108.
   treatment, 111.
Endocervicitis, acute septic, 20.
   gonorrheal, acute, 25.
   diagnosis, 25.
   symptoms, 25.

INDEX.

Endocervicitis, gonorrheal, acute, treatment, 25.
  gonorrheal, chronic, 25.
    symptoms, 26.
    treatment, 26.
  gonorrheal, latent. See *Endocervicitis, gonorrheal, chronic.*
  septic, diagnosis, 22.
    symptoms, 20.
    treatment, 24.
Endometritis, 17.
  general considerations, 17.
  gonorrheal, acute, 47.
    diagnosis, 48.
    sequelæ, 49.
    symptoms, 47.
    treatment, non-operative, 48.
    treatment, operative, 49.
  gonorrheal, chronic, 50.
    diagnosis, 51.
    symptoms, 50.
    treatment, 51.
  non-virulent, 18.
  puerperal septic, 35.
    cul-de-sac operation in, 44.
    curettage in, 43.
    irrigation in, 42.
    septic, treatment, 41.
  virulent, 18.
  septic, 26.
    acute, 26.
    acute, differential diagnosis, 30.
    sequelæ, 30.
    symptoms, 26.
    treatment, 30
  tubercular, 52.
    diagnosis, 52.
    sequelæ, 53.

Endometritis, tubercular, symptoms, 52.
  treatment, 52.
Exploratory vaginal section, 136.
  advantages of, over abdominal method, 144.
Exposure as a cause of pelvic peritonitis, 96.

F

Fluids in pelvic inflammation, 61.
Forceps, French traction, 234.
  Hunter's, 235.
  Luer's polypus, 234.
  Pryor's, 235.
  removal of, after hysterectomy, 221.
  Sims' artery, 235.
Formulæ, 237.
French traction forceps, 234.
Fritsch-Bozeman double-current irrigating tubes, 135.

G

Gauze, iodoform, 237.
  sterilization of, 240.
Genital sclerosis, 26.
Gonococci as a cause of pelvic peritonitis, 92.
Gonorrhea a cause of pelvic peritonitis, 92.
Gonorrheal endocervicitis, acute, 25.
  endocervicitis, chronic, 25. See *Endocervicitis.*
  endocervicitis, latent, 25.
  endometritis, acute, 47. See *Endometritis.*
  chronic, 50. See *Endometritis.*
  peritonitis, 92.

INDEX.

Gonorrheal salpingitis, acute, 63.
 See also *Salpingitis*.
 chronic, 72. See also *Salpingitis*.

H

Hand-basins, sterilization of, 240.
Heart in pelvic peritonitis, 101.
Hemorrhage, secondary, after vaginal ablation, 227.
Hemostasis in vaginal ablation of uterus, 165.
Hemostatic forceps, 235.
Hernia after hysterectomy, 166.
Hunter's sponge forceps, 235.
Hydrosalpinx, 80.
 conservative operation for, 157.
 treatment, 80.
Hysterectomy, vaginal, 163. See *Vaginal Ablation*.
 after treatment of, 219.
 anodynes in after-treatment of, 221
 behavior of wound after, 222.
 removal of forceps after, 221.
 vagino-abdominal, in puerperal state, 215.

I

Infection, puerperal, 34.
Inflammation, intra-uterine. See *Endometritis*.
 of ovaries, 106.
 pelvic, 54. See *Pelvic Inflammation*.
Instruments, 231.
 sterilization of, 239.
Intestinal paralysis after vaginal ablation, 226.
Intra-uterine catheters, 234.
Intra-venous injection of normal salt solution, 229.

Iodoform gauze, 237.
Irrigating fluids, 239.
Irrigation of uterus after curettage, 129.
 puerperal endometritis, 42.
Irrigations, repeated, 133.

J

Jackson's speculum, 134.

K

Kangaroo tendon, sterilization of, 240.
Kidneys in pelvic peritonitis, 101.

L

Latent gonorrheal endocervicitis, 25. See *Endocervicitis*.
Local applications in pelvic inflammation, 62.
Luer's polypus forceps, 234.
Lungs in pelvic peritonitis, 101.
Lysol, 237.

M

Mikulicz pelvic dressing, 202.
Morcellation, 209.

N

Nephritis after vaginal ablation, 225.
Non-purulent endometritis, 18.
Normal salt solution, 237.
 intra-venous injection of, 229.
 subcutaneous injection of, 230.
Nutrient enema, 238.

O

Occluded tubes, conservative treatment, 159.
Operating table, 236.

INDEX. 245

Operator, sterilization of, 238.
Opiates in pelvic inflammation, 60.
Ovarian abscess, 108.
　symptoms, 110.
　treatment, 113.
　apoplexy, 107.
　　conservative treatment of, 159.
　　treatment, 112.
　sclerosis, 107.
　　treatment, 111.
Ovaries, adherent, conservative treatment, 159.
　cystic, 107.
　　conservative treatment, 158.
　　treatment, 111.
　cystic degeneration of, 107.
　inflammatory diseases of, 106.
Ovaritis, acute, 106.
　edematous, 108.
　　treatment, 111.
　symptoms, 109.

P

Pachysalpingitis, 72, 74.
　treatment of, 79.
Packing applicator, Pryor's, 135.
Packing of uterus after curetting, 129.
Pain in pelvic peritonitis, 99.
Péan retractor, 140, 231, 232.
　long retractor, 227.
Pelvic inflammation, 54.
　abdominal dressings in, 61.
　diagnosis, 57.
　diagnosis from appendicitis, 57.
　diagnosis from cystitis, 58.
　diagnosis from general suppurative peritonitis, 59.

Pelvic inflammation, diagnosis from suppurating ovarian cyst, 58.
　diagnosis from ureteritis, 58.
　diet in, 62.
　fluids in, 61.
　douches in, 61.
　intestinal cleanliness in, 60.
　local applications in, 62.
　opiates in, 60.
　treatment, 62.
Mikulicz dressing, 202.
peritonitis, 90.
　analgesia in, 99.
　causes, 91.
　chills in, 101.
　colon bacillus in, 91.
　diagnosis, 102.
　digestive symptoms in, 101.
　gonococci in, 92.
　heart in, 101.
　kidneys in, 101.
　lungs in, 101.
　pain in, 99.
　prognosis, 102.
　pulse in, 100.
　staphylococci in, 92.
　streptococci in, 94.
　suppurative, treatment, 103.
　symptoms, 96.
　temperature in, 100.
　treatment, 103.
　tubercular, 104.
　　symptoms, 105.
　　treatment, 105.
　tympanites in, 99.
suppuration, diffuse, 123.
　symptoms, 124.
　treatment, 124.
Peri-ovaritis, acute, 106.
　symptoms, 110.

INDEX.

Peri-ovaritis, treatment, 110.
Peritonitis, pelvic, 90.
    causes, 91.
    diagnosis, 102.
    prognosis, 102.
    symptoms, 96.
    treatment, 103.
    tubercular, 104.
        symptoms, 105.
        treatment, 105.
    primary purulent, 94.
Pneumonia after vaginal ablation, 225.
Polypus of cervix uteri, 22.
    treatment, 24.
Pryor-Péan trowel, 231, 233.
Pryor's blunt bullet forceps, 134.
    forceps, 235.
    intra-uterine traction forceps, 231.
    operating table, 236.
    packing applicator, 135.
    retracting grooved director, 194.
    trowel, 233.
    uterine dilator, 135.
Puerperal endometritis, septic, 35. See *Endometritis, puerperal*.
    infection, 34.
    state, vagino-abdominal hysterectomy in, 215.
Pulse in pelvic peritonitis, 99.
Purulent endometritis, 18.
Pyosalpinx, 80.
    diagnosis, 87.
    sequelæ, 88.
    symptoms, 83.
    treatment, 88.

Q

Quinin solution, 237.

R

Repeated irrigations after curettage of uterus, 133.
Retractors, 231.
    Péan's, 227.
Retroposition, adherent, 117.
    operation for, 118.
Rubber goods, sterilization of, 239.

S

Salpingitis, 63.
    gonorrheal, acute, 63.
        diagnosis, 67.
        sequelæ, 68.
        symptoms, 65.
        treatment, 67.
    septic, acute, 69.
        treatment, 71.
    tubercular, 89.
        symptoms, 90.
        treatment, 90.
Salpingo-oöphoritis, acute, conservative operation for, 155.
Salpingostomy, 158.
Sclerosis, genital, 26, 74.
    treatment of, 74.
    ovarian, 107.
        treatment, 111.
    tubal, 63.
Secondary hemorrhage after vaginal ablation, 227.
Section, vaginal, preparation of patient for, 161.
Speculum, Jackson, 134.
Ségond's incisions, 179.
Septic endocervicitis, 20.
    diagnosis, 22.
    symptoms, 20.
    treatment, 24.

INDEX. 247

Septic endometritis, 26. See *Endometritis*.
    puerperal endometritis, 35.
    treatment, 41.
    salpingitis, acute, 69. See also *Salpingitis*.
    chronic, 72. See also *Salpingitis*.
Sheets, sterilization of, 240.
Silks, sterilization of, 240.
Silkworm gut, sterilization of, 240.
Silver wire, sterilization of, 240.
Sims' artery forceps, 235.
    curettes, 135.
    needle-holder, 236.
    tampon screw, 135.
Staphylococci, as a cause of pelvic peritonitis, 92.
Sterilization, 238.
Streptococci as a cause of pelvic peritonitis, 94.
Subcutaneous injection of normal salt solution, 230.
Suppuration, diffuse pelvic, 123.
    symptoms, 124.
    treatment, 124.
Suppurative pelvic peritonitis, treatment, 103.
Sutures in hysterectomy, 166.
    sterilization of, 240.

T

Tampon screw, Sims', 135.
Temperature in pelvic peritonitis, 100.
Thiersch solution, 237.
Towels, sterilization of, 240.
Traction instruments, 231.
Transfusion fluid, sterilization of, 239.

Trauma as a cause of pelvic peritonitis, 96.
Treatment, conservative, 146. See *Conservative treatment*.
Tubal sclerosis, 63.
Tubercular endometritis, 52. See *Endometritis*.
    pelvic peritonitis, 104.
    symptoms, 105.
    treatment, 105.
    salpingitis, 89.
Tympanites in pelvic peritonitis, 99.

U

Ureter, wounds of, in vaginal ablation, 224.
Ureteritis, diagnosis of pelvic inflammation from, 58.
Uterine colic in gonorrheal endometritis, 48.
    dilator, Pryor's, 135.
Uterus, ablation of, en masse, 180.
    by hemisection, 187.
    curettage of, 126.
    morcellation of, 209.
    packing of, after curetting, 129.
    vaginal ablation of, 163. See also *Vaginal ablation of uterus*, 163.

V

Vaginal ablation of uterus, 163.
    accidents and complications, 223.
    accidents in, 166.
    after-treatment, 219.
    convalescence from, 167.
    drainage in, 165.
    enucleation in, 165.
    general considerations, 163.

Vaginal ablation of uterus, hemostasis in, 165.
   hernia after, 166.
   instruments for, 166.
   operation, 168.
   posture for, 168.
   results of, 167.
   secondary hemorrhage after, 227.
   separation of adhesions in, 163.

Vaginal ablation of uterus, hysterectomy, 163. See *Vaginal ablation*.
   sutures in, 166.
   section, exploratory, 136.
   preparation of patient for, 161.
Vagino-abdominal hysterectomy in puerperal state, 215.

W

Wales bougie, 236.
Womb cramps, 27.

Medical and Surgical Works

PUBLISHED BY

W. B. SAUNDERS, 925 Walnut Street, Philadelphia, Pa.

	PAGE
Abbott on Transmissible Diseases	18
American Pocket Medical Dictionary	35
*American Text-Book of Applied Therapeutics	8
*American Text-Book of Dis. of Children	13
*An American Text-Book of Diseases of the Eye, Ear, Nose, and Throat	15
*An American Text-Book of Genito-Urinary and Skin Diseases	14
*American Text-Book of Gynecology	12
*American Text-Book of Legal Medicine	44
*American Text-Book of Obstetrics	9
*American Text-Book of Pathology	44
*American Text-Book of Physiology	7
*American Text-Book of Practice	10
*American Text-Book of Surgery	11
Anders' Theory and Practice of Medicine	21
Ashton's Obstetrics	43
Atlas of Skin Diseases	28
Ball's Bacteriology	43
Bastin's Laboratory Exercises in Botany	36
Beck's Surgical Asepsis	41
Boislinière's Obstetric Accidents	39
Brockway's Physics	43
Burr's Nervous Diseases	41
Butler's Materia Medica and Therapeutics	24
Cerna's Notes on the Newer Remedies	32
Chapin's Compendium of Insanity	35
Chapman's Medical Jurisprudence	41
Church and Peterson's Nervous and Mental Diseases	17
Clarkson's Histology	33
Cohen and Eshner's Diagnosis	43
Corwin's Diagnosis of the Thorax	37
Cragin's Gynæcology	43
Crookshank's Text-Book of Bacteriology	27
DaCosta's Manual of Surgery	23
De Schweinitz's Diseases of the Eye	29
Dorland's Pocket Medical Dictionary	35
Dorland's Obstetrics	41
Frothingham's Bacteriological Guide	30
Garrigues' Diseases of Women	34
Gleason's Diseases of the Ear	43
*Gould and Pyle's Curiosities of Medicine	17
Grafstrom's Massage	28
Griffith's Care of the Baby	38
Griffith's Infant's Weight Chart	39
Gross's Autobiography	26
Hampton's Nursing	39
Hare's Physiology	43
Hart's Diet in Sickness and in Health	36
Haynes' Manual of Anatomy	41
Heisler's Embryology	19
Hirst's Obstetrics	20
Hyde's Syphilis and Venereal Diseases	41
International Text-Book of Surgery	6
Jackson's Diseases of the Eye	19
Jackson and Gleason's Diseases of the Eye, Nose, and Throat	43
Keating's Pronouncing Dictionary	26
Keating's Life Insurance	39
Keen's Operation Blanks	36
Keen's Surgery of Typhoid Fever	22

	PAGE
Kyle's Diseases of Nose and Throat	18
Lainé's Temperature Charts	32
Levy & Klemperer's Clinical Bacteriology	44
Lockwood's Practice of Medicine	41
Long's Syllabus of Gynecology	34
Macdonald's Surgical Diagnosis and Treatment	22
McFarland's Pathogenic Bacteria	30
Mallory and Wright's Pathological Technique	22
Martin's Surgery	43
Martin's Minor Surgery, Bandaging, and Venereal Diseases	43
Meigs' Feeding in Early Infancy	30
Moore's Orthopedic Surgery	23
Morris' Materia Medica and Therapeutics	43
Morris' Practice of Medicine	43
Morten's Nurses' Dictionary	38
Nancrede's Anatomy and Dissection	31
Nancrede's Anatomy	43
Nancrede's Principles of Surgery	19
Norris' Syllabus of Obstetrical Lectures	37
Penrose's Diseases of Women	24
Powell's Diseases of Children	43
Pryor's Pelvic Inflammations	33
Pye's Bandaging and Surgical Dressing	23
Raymond's Physiology	41
Saundby's Renal and Urinary Diseases	25
*Saunders' American Year-Book of Medicine and Surgery	16
Saunders' Medical Hand-Atlases	3, 4, 5
Saunders' Pocket Medical Formulary	35
Saunders' New Series of Manuals	40, 41
Saunders' Series of Question Compends	42, 43
Sayre's Practice of Pharmacy	43
Semple's Pathology and Morbid Anatomy	43
Semple's Legal Medicine and Toxicology	43
Senn's Genito-Urinary Tuberculosis	24
Senn's Tumors	25
Senn's Syllabus of Lectures on Surgery	37
Shaw's Nervous Diseases and Insanity	43
Starr's Diet-Lists for Children	38
Stelwagon's Diseases of the Skin	43
Stengel's Pathology	7
Stevens' Materia Medica and Therapeutics	32
Stevens' Practice of Medicine	31
Stewart's Manual of Physiology	37
Stewart and Lawrance's Medical Electricity	43
Stoney's Materia Medica for Nurses	31
Stoney's Practical Points in Nursing	27
Sutton and Giles' Diseases of Women	29, 41
Thomas's Diet-List and Sick-Room	38
Thornton's Dose-Book and Manual of Prescription-Writing	41
Van Valzah and Nisbet's Diseases of the Stomach	21
Vecki's Sexual Impotence	33
Vierordt and Stuart's Medical Diagnosis	28
Warren's Surgical Pathology	25
Watson's Handbook for Nurses	26
Wolff's Chemistry	43
Wolff's Examination of Urine	43

GENERAL INFORMATION.

One Price. One price absolutely without deviation. No discounts allowed, regardless of the number of books purchased at one time. Prices on all works have been fixed extremely low, with the view to selling them strictly net and for cash.

Orders. An order accompanied by remittance will receive prompt attention, books being sent to any address in the United States, by mail or express, all charges prepaid. We prefer to send books **by express** when possible.

Cash or Credit. To physicians of approved credit who furnish satisfactory references our books will be sent free of C. O. D. **One volume or two** on thirty days' time if credit is desired; larger purchases on monthly payment plan. See offer below.

How to Send Money by Mail. There are four ways by which money can be sent at our risk, namely: a post-office money order, an express money order, a bank-check (draft), and in a registered letter. Money sent in any other way is at the sender's risk. Silver should not be sent through the mail.

Shipments. All books, being packed in patent metal-edged boxes, necessarily reach our patrons by mail or express in excellent condition.

Subscription Books. Books in this catalogue marked with a star (*) are for sale by subscription only, and may be secured by ordering them through any of our authorized travelling salesmen, or direct from the Philadelphia office: they are **not** for sale by booksellers. All other books in our catalogue can be procured of any bookseller at the advertised price, or directly from us.

Miscellaneous Books. We carry in stock only our own publications, but can supply the publications of other houses (except subscription books) on receipt of publisher's price.

Latest Editions. In every instance the latest revised edition is sent.

Bindings. In ordering, be careful to state the style of binding desired—Cloth, Sheep, or Half Morocco.

Special Offer. Monthly Payment Plan. To physicians of approved credit who furnish satisfactory references books will be sent express prepaid; terms, $5.00 cash upon delivery of books, and monthly payments of $5.00 thereafter until full amount is paid. Any of the publications of W. B. Saunders (100 titles to select from) may be had in this way at catalogue price, including the American Text-Book Series, the Medical Hand-Atlases, etc. All payments to be made by mail or otherwise, free of all expense to us.

SAUNDERS'
MEDICAL HAND-ATLASES.

THE series of books included under this title consists of authorized translations into English of the world-famous **Lehmann Medicinische Handatlanten**, which for **scientific accuracy, pictorial beauty, compactness,** and **cheapness** surpass any similar volumes ever published. Each volume contains from **50 to 100 colored plates**, executed by the most skilful German lithographers, besides numerous illustrations in the text. There is a full and appropriate **description**, and each book contains a condensed but adequate **outline of the subject** to which it is devoted.

In planning this series arrangements were made with representative publishers in the chief medical centers of the world for the publication of translations of the atlases into nine different languages, the lithographic plates for all being made in Germany, where work of this kind has been brought to the greatest perfection. The enormous expense of making the plates being shared by the various publishers, the cost to each one was reduced to practically one-tenth. Thus by reason of their **universal translation** and reproduction, affording international distribution, the publishers have been enabled to secure for these atlases the **best artistic and professional talent,** to produce them in the most **elegant style,** and yet to offer them at a **price heretofore unapproached in cheapness.** The great success of the undertaking is demonstrated by the fact that the volumes have already appeared in **thirteen different languages** —German, English, French, Italian, Russian, Spanish, Japanese, Dutch, Danish, Swedish, Roumanian, Bohemian, and Hungarian.

In view of the unprecedented success of these works, Mr. Saunders has contracted with the publisher of the original German edition for **one hundred thousand copies** of the atlases. In consideration of this enormous undertaking, the publisher has been enabled to prepare and furnish special additional colored plates, making the series even **handsomer and more complete** than was originally intended.

As an indication of the great practical value of the atlases and of the immense favor with which they have been received, it should be noted that the **Medical Department of the U. S. Army** has adopted the "Atlas of Operative Surgery," as its standard, and has ordered the book in large quantities for distribution to the various regiments and army posts.

The same careful and competent editorial **supervision** has been secured in the English edition as in the originals. The translations have been edited by the **leading American specialists** in the different subjects.

(For List of Volumes in this Series, see next two pages.)

SAUNDERS' MEDICAL HAND-ATLASES.

VOLUMES NOW READY.

Atlas and Epitome of Internal Medicine and Clinical Diagnosis. By Dr. CHR. JAKOB, of Erlangen. Edited by AUGUSTUS A. ESHNER, M. D., Professor of Clinical Medicine, Philadelphia Polyclinic. With 68 colored plates, 64 text-illustrations, and 259 pages of text. Cloth, $3.00 net.

"The charm of the book is its clearness, conciseness, and the accuracy and beauty of its illustrations. It deals with facts. It vividly illustrates those facts. It is a scientific work put together for ready reference."—*Brooklyn Medical Journal.*

Atlas of Legal Medicine. By DR. E. R. VON HOFMANN, of Vienna. Edited by FREDERICK PETERSON, M. D., Chief of Clinic, Nervous Dept., College of Physicians and Surgeons, New York. With 120 colored figures on 56 plates, and 193 beautiful half-tone illustrations. Cloth, $3.50 net.

"Hofmann's 'Atlas of Legal Medicine' is a unique work. This immense field finds in this book a pictorial presentation that far excels anything with which we are familiar in any other work."—*Philadelphia Medical Journal.*

Atlas and Epitome of Diseases of the Larynx. By DR. L. GRUNWALD, of Munich. Edited by CHARLES P. GRAYSON, M. D., Physician-in-Charge, Throat and Nose Department, Hospital of the University of Pennsylvania. With 107 colored figures on 44 plates, 25 text-illustrations, and 103 pages of text. Cloth, $2.50 net.

"Aided as it is by magnificently executed illustrations in color, it cannot fail of being of the greatest advantage to students, general practitioners, and expert laryngologists."—*St. Louis Medical and Surgical Journal.*

Atlas and Epitome of Operative Surgery. By DR. O. ZUCKERKANDL, of Vienna. Edited by J. CHALMERS DACOSTA, M. D., Professor of Practice of Surgery and Clinical Surgery, Jefferson Medical College, Philadelphia. With 24 colored plates, 217 text-illustrations, and 395 pages of text. Cloth, $3.00 net.

"We know of no other work that combines such a wealth of beautiful illustrations with clearness and conciseness of language, that is so entirely abreast of the latest achievements, and so useful both for the beginner and for one who wishes to increase his knowledge of operative surgery."—*Münchener medicinische Wochenschrift.*

Atlas and Epitome of Syphilis and the Venereal Diseases. By PROF. DR. FRANZ MRACEK, of Vienna. Edited by L. BOLTON BANGS, M. D., Professor of Genito-Urinary Surgery, University and Bellevue Hospital Medical College, New York. With 71 colored plates, 16 black-and-white illustrations, and 122 pages of text. Cloth, $3.50 net.

"A glance through the book is almost like actual attendance upon a famous clinic."—*Journal of the American Medical Association.*

Atlas and Epitome of External Diseases of the Eye. By DR. O. HAAB, of Zurich. Edited by G. E. DE SCHWEINITZ, M. D., Professor of Ophthalmology, Jefferson Medical College, Philadelphia. With 76 colored illustrations on 40 plates, and 228 pages of text. Cloth, $3.00 net.

"It is always difficult to represent pathological appearances in colored plates, but this work seems to have overcome these difficulties, and the plates, with one or two exceptions, are absolutely satisfactory."—*Boston Medical and Surgical Journal.*

Atlas and Epitome of Skin Diseases. By PROF. DR. FRANZ MRACEK, of Vienna. Edited by HENRY W. STELWAGON, M. D., Clinical Professor of Dermatology, Jefferson Medical College, Philadelphia. With 63 colored plates, 39 half-tone illustrations, and 200 pages of text. Cloth, $3.50 net.

"The importance of personal inspection of cases in the study of cutaneous diseases is readily appreciated, and next to the living subjects are pictures which will show the appearance of the disease under consideration. Altogether the work will be found of very great value to the general practitioner."—*Journal of the American Medical Association.*

SAUNDERS' MEDICAL HAND-ATLASES.

VOLUMES IN PRESS FOR EARLY PUBLICATION.

Atlas and Epitome of Diseases Caused by Accidents. By DR. ED. GOLEBIEWSKI, of Berlin. Translated and edited with additions by PEARCE BAILEY, M.D., Attending Physician to the Department of Corrections and to the Almshouse and Incurable Hospitals, New York. With 40 colored plates, 143 text-illustrations, and 600 pages of text.

Atlas and Epitome of Special Pathological Histology. By DR. H. DÜRCK, of Munich. Edited by LUDVIG HEKTOEN, M.D., Professor of Pathology, Rush Medical College, Chicago. Two volumes, with about 120 colored plates, numerous text-illustrations, and copious text.

Atlas and Epitome of General Pathological Histology. With an Appendix on Patho-histological Technic. By DR. H. DÜRCK, of Munich. Edited by LUDVIG HEKTOEN, M.D., Professor of Pathology, Rush Medical College, Chicago. With 80 colored plates, numerous text-illustrations, and copious text.

Atlas and Epitome of Gynecology. By DR. O. SCHÄFFER, of the University of Heidelberg. With 90 colored plates, 65 text-illustrations, and 308 pages of text. Edited by RICHARD C. NORRIS, A. M., M.D., Gynecologist to the Philadelphia and the Methodist Episcopal Hospitals.

IN PREPARATION.

Atlas and Epitome of Orthopedic Surgery. By DR. SCHULTESS and DR. LÜNING, of Zurich. About 100 colored illustrations.

Atlas and Epitome of Operative Gynecology. By DR. O. SCHÄFFER, of Heidelberg. With 40 colored plates and numerous illustrations in black and white from original paintings.

Atlas and Epitome of Diseases of the Ear. Edited by PROF. DR. POLITZER, of Vienna, and DR. G. BRÜHL, of Berlin. With 120 colored illustrations and about 200 pages of text.

Atlas and Epitome of General Surgery. Edited by DR. MARWEDEL, with the coöperation of PROF. DR. CZERNY. With about 200 colored illustrations.

Atlas and Epitome of Psychiatry. By DR. WILH. WEYGANDT, of Würzburg. With about 120 colored illustrations.

Atlas and Epitome of Normal Histology. By DR. JOHANNES SOBOTTA, of Würzburg. With 80 colored plates and numerous illustrations.

Atlas and Epitome of Topographical Anatomy. By PROF. DR. SCHULTZE, of Würzburg. About 100 colored illustrations and a very copious text.

*THE INTERNATIONAL TEXT-BOOK OF SURGERY. In two volumes. By American and British authors. Edited by J. COLLINS WARREN, M.D., LL.D., Professor of Surgery, Harvard Medical School, Boston; Surgeon to the Massachusetts General Hospital; and A. PEARCE GOULD, M. S., F. R. C. S., Eng., Lecturer on Practical Surgery and Teacher of Operative Surgery, Middlesex Hospital Medical School; Surgeon to the Middlesex Hospital, London, England. Vol. I.—**General and Operative Surgery.**—Handsome octavo volume of 947 pages, with 458 beautiful illustrations, and 9 lithographic plates. Vol. II.—**Special or Regional Surgery.**—Handsome octavo volume of 1050 pages, with over 500 woodcuts and half-tones, and 8 lithographic plates. Prices per volume: Cloth, $5.00 net; Half-Morocco, $6.00 net.

Just Issued.

In presenting a new work on surgery to the medical profession the publisher feels that he need offer no apology for making an addition to the list of excellent works already in existence. Modern surgery is still in the transition stage of its development. The art and science of surgery are advancing rapidly, and the number of workers is now so great and so widely spread through the whole of the civilized world that there is certainly room for another work of reference which shall be untrammelled by many of the traditions of the past, and shall at the same time present with due discrimination the results of modern progress. There is a real need among practitioners and advanced students for a work on surgery encyclopedic in scope, yet so condensed in style and arrangement that the matter usually diffused through four or five volumes shall be given in one-half the space and at a correspondingly moderate cost.

The ever-widening-field of surgery has been developed largely by special work, and this method of progress has made it practically impossible for one man to write authoritatively on the vast range of subjects embraced in a modern text-book of surgery. In order, therefore, to accomplish their object, the editors have sought the aid of men of wide experience and established reputation in the various departments of surgery.

CONTRIBUTORS:

Dr. Robert W. Abbe.
C. H. Golding Bird.
E. H. Bradford.
W. T. Bull.
T. G. A. Burns.
Herbert L. Burrell.
R. C. Cabot.
I. H. Cameron.
James Cantlie.
W. Watson Cheyne.
William B. Clarke.
William B. Coley.
Edw. Treacher Collins.
H. Holbrook Curtis.
J. Chalmers Da Costa.
N. P. Dandridge.
John B. Deaver.
J. W. Elliot.
Harold Ernst.

Dr. Christian Fenger.
W. H. Forwood.
George R. Fowler.
George W. Gay.
A. Pearce Gould.
J. Orne Green.
John B. Hamilton.
M. L. Harris.
Fernand Henrotin.
G. H. Makins.
Rudolph Matas.
Charles McBurney.
A. J. McCosh.
L. S. McMurtry.
J. Ewing Mears.
George H. Monks.
John Murray.
Robert W. Parker.

Dr. Rushton Parker.
George A. Peters.
Franz Pfaff.
Lewis S. Pilcher.
James J. Putnam.
M. H. Richardson.
A. W. Mayo Robson.
W. L. Rodman.
C. A. Siegfried.
G. B. Smith.
W. G. Spencer.
J. Bland Sutton.
L. McLane Tiffany.
H. Tuholske.
Weller Van Hook.
James P. Warbasse.
J. Collins Warren.
De Forest Willard.

*AN AMERICAN TEXT-BOOK OF PHYSIOLOGY. Edited by WILLIAM H. HOWELL, PH. D., M. D., Professor of Physiology in the Johns Hopkins University, Baltimore, Md. One handsome octavo volume of 1052 pages, fully illustrated. Prices: Cloth, $6.00 net; Sheep or Half-Morocco, $7.00 net.

This work is the most notable attempt yet made in America to combine in one volume the entire subject of Human Physiology by well-known teachers who have given especial study to that part of the subject upon which they write. The completed work represents the present status of the science of Physiology, particularly from the standpoint of the student of medicine and of the medical practitioner.

The collaboration of several teachers in the preparation of an elementary text-book of physiology is unusual, the almost invariable rule heretofore having been for a single author to write the entire book. One of the advantages to be derived from this collaboration method is that the more limited literature necessary for consultation by each author has enabled him to base his elementary account upon a comprehensive knowledge of the subject assigned to him; another, and perhaps the most important, advantage is that the student gains the point of view of a number of teachers. In a measure he reaps the same benefit as would be obtained by following courses of instruction under different teachers. The different standpoints assumed, and the differences in emphasis laid upon the various lines of procedure, chemical, physical, and anatomical, should give the student a better insight into the methods of the science as it exists to-day. The work will also be found useful to many medical practitioners who may wish to keep in touch with the development of modern physiology.

CONTRIBUTORS:

HENRY P. BOWDITCH, M. D.,
Professor of Physiology, Harvard Medical School.

JOHN G. CURTIS, M. D.,
Professor of Physiology, Columbia University, N. Y. (College of Physicians and Surgeons).

HENRY H. DONALDSON, Ph. D.,
Head-Professor of Neurology, University of Chicago.

W. H. HOWELL, Ph. D., M. D.,
Professor of Physiology, Johns Hopkins University.

FREDERIC S. LEE, Ph. D.,
Adjunct Professor of Physiology, Columbia University, N. Y. (College of Physicians and Surgeons).

WARREN P. LOMBARD, M. D.,
Professor of Physiology, University of Michigan.

GRAHAM LUSK, Ph. D.,
Professor of Physiology, Yale Medical School.

W. T. PORTER, M. D.,
Assistant Professor of Physiology, Harvard Medical School.

EDWARD T. REICHERT, M. D.,
Professor of Physiology, University of Pennsylvania.

HENRY SEWALL, Ph. D., M. D.,
Professor of Physiology, Medical Department, University of Denver.

"We can commend it most heartily, not only to all students of physiology, but to every physician and pathologist, as a valuable and comprehensive work of reference, written by men who are of eminent authority in their own special subjects."—*London Lancet.*

"To the practitioner of medicine and to the advanced student this volume constitutes, we believe, the best exposition of the present status of the science of physiology in the English language."—*American Journal of the Medical Sciences.*

*AN AMERICAN TEXT-BOOK OF APPLIED THERAPEU-TICS. For the Use of Practitioners and Students. Edited by JAMES C. WILSON, M. D., Professor of the Practice of Medicine and of Clinical Medicine in the Jefferson Medical College. One handsome octavo volume of 1326 pages. Illustrated. Prices: Cloth, $7.00 net; Sheep or Half-Morocco, $8.00 net.

The *arrangement* of this volume has been based, so far as possible, upon modern pathologic doctrines, beginning with the intoxications, and following with infections, diseases due to internal parasites, diseases of undetermined origin, and finally the disorders of the several bodily systems—digestive, respiratory, circulatory, renal, nervous, and cutaneous. It was thought proper to include also a consideration of the disorders of pregnancy.

The articles, with two exceptions, are the contributions of American writers. Written from the standpoint of the practitioner, the aim of the work is to facilitate the application of knowledge to the prevention, the cure, and the alleviation of disease. The endeavor throughout has been to conform to the title of the book—Applied Therapeutics—to indicate the course of treatment to be pursued at the bedside, rather than to name a list of drugs that have been used at one time or another.

The list of contributors comprises the names of many who have acquired distinction as practitioners and teachers of practice, of clinical medicine, and of the specialties.

CONTRIBUTORS:

Dr. I. E. Atkinson, Baltimore, Md.
Sanger Brown, Chicago, Ill.
John B. Chapin, Philadelphia, Pa.
William C. Dabney, Charlottesville, Va.
John Chalmers DaCosta, Philada., Pa.
I. N. Danforth, Chicago, Ill.
John L. Dawson, Jr., Charleston, S. C.
F. X. Dercum, Philadelphia, Pa.
George Dock, Ann Arbor, Mich.
Robert T. Edes, Jamaica Plain, Mass.
Augustus A. Eshner, Philadelphia, Pa.
J. T. Eskridge, Denver, Col.
F. Forchheimer, Cincinnati, O.
Carl Frese, Philadelphia, Pa.
Edwin E. Graham, Philadelphia, Pa.
John Guiteras, Philadelphia, Pa.
Frederick P. Henry, Philadelphia, Pa.
Guy Hinsdale, Philadelphia, Pa.
Orville Horwitz, Philadelphia, Pa.
W. W. Johnston, Washington, D. C.
Ernest Laplace, Philadelphia, Pa.
A. Laveran, Paris, France.

Dr. James Hendrie Lloyd, Philadelphia, Pa.
John Noland Mackenzie, Baltimore, Md.
J. W. McLaughlin, Austin, Texas.
A. Lawrence Mason, Boston, Mass.
Charles K. Mills, Philadelphia, Pa.
John K. Mitchell, Philadelphia, Pa.
W. P. Northrup, New York City.
William Osler, Baltimore, Md.
Frederick A. Packard, Philadelphia, Pa.
Theophilus Parvin, Philadelphia, Pa.
Benven Rake, London, England.
E. O. Shakespeare, Philadelphia, Pa.
Wharton Sinkler, Philadelphia, Pa.
Louis Starr, Philadelphia, Pa.
Henry W. Stelwagon, Philadelphia, Pa.
James Stewart, Montreal, Canada.
Charles G. Stockton, Buffalo, N. Y.
James Tyson, Philadelphia, Pa.
Victor C. Vaughan, Ann Arbor, Mich.
James T. Whittaker, Cincinnati, O.
J. C. Wilson, Philadelphia, Pa.

"As a work either for study or reference it will be of great value to the practitioner, as it is virtually an exposition of such clinical therapeutics as experience has taught to be of the most value. Taking it all in all, no recent publication on therapeutics can be compared with this one in practical value to the working physician."—*Chicago Clinical Review.*

"The whole field of medicine has been well covered. The work is thoroughly practical, and while it is intended for practitioners and students, it is a better book for the general practitioner than for the student. The young practitioner especially will find it extremely suggestive and helpful."—*The Indian Lancet.*

*AN AMERICAN TEXT-BOOK OF OBSTETRICS. Edited by RICHARD C. NORRIS, M. D.; Art Editor, ROBERT L. DICKINSON, M. D. One handsome octavo volume of over 1000 pages, with nearly 900 colored and half-tone illustrations. Prices: Cloth, $7.00 net; Sheep or Half Morocco, $8.00 net.

The advent of each successive volume of the *series* of the AMERICAN TEXT-BOOKS has been signalized by the most flattering comment from both the Press and the Profession. The high consideration received by these text-books, and their attainment to an authoritative position in current medical literature, have been matters of deep *international* interest, which finds its fullest expression in the demand for these publications from all parts of the civilized world.

In the preparation of the "AMERICAN TEXT-BOOK OF OBSTETRICS" the editor has called to his aid proficient collaborators whose professional prominence entitles them to recognition, and whose disquisitions exemplify **Practical Obstetrics.** While these writers were each assigned special themes for discussion, the correlation of the subject-matter is, nevertheless, such as ensures logical connection in treatment, the deductions of which thoroughly represent the latest advances in the science, and which elucidate *the best modern methods of procedure.*

The more conspicuous feature of the treatise is its wealth of illustrative matter. The production of the illustrations had been in progress for several years, under the personal supervision of Robert L. Dickinson, M. D., to whose artistic judgment and professional experience is due the **most sumptuously illustrated work of the period.** By means of the photographic art, combined with the skill of the artist and draughtsman, conventional illustration is superseded by rational methods of delineation.

Furthermore, the volume is a revelation as to the possibilities that may be reached in mechanical execution, through the unsparing hand of its publisher.

CONTRIBUTORS:

Dr. James C. Cameron.
Edward P. Davis.
Robert L. Dickinson.
Charles Warrington Earle.
James H. Etheridge.
Henry J. Garrigues.
Barton Cooke Hirst.
Charles Jewett.

Dr. Howard A. Kelly.
Richard C. Norris.
Chauncey D. Palmer.
Theophilus Parvin.
George A. Piersol.
Edward Reynolds.
Henry Schwarz.

"At first glance we are overwhelmed by the magnitude of this work in several respects, viz.: First, by the size of the volume, then by the array of eminent teachers in this department who have taken part in its production, then by the profuseness and character of the illustrations, and last, but not least, the conciseness and clearness with which the text is rendered. This is an entirely new composition, embodying the highest knowledge of the art as it stands to-day by authors who occupy the front rank in their specialty, and there are many of them. We cannot turn over these pages without being struck by the superb illustrations which adorn so many of them. We are confident that this most practical work will find instant appreciation by practitioners as well as students."—*New York Medical Times.*

Permit me to say that your American Text-Book of Obstetrics is the most magnificent medical work that I have ever seen. I congratulate you and thank you for this superb work, which alone is sufficient to place you first in the ranks of medical publishers.

With profound respect I am sincerely yours, ALEX. J. C. SKENE.

* **AN AMERICAN TEXT-BOOK OF THE THEORY AND PRACTICE OF MEDICINE. By American Teachers.** Edited by WILLIAM PEPPER, M. D., LL.D., Provost and Professor of the Theory and Practice of Medicine and of Clinical Medicine in the University of Pennsylvania. Complete in two handsome royal-octavo volumes of about 1000 pages each, with illustrations to elucidate the text wherever necessary. Price per Volume: Cloth, $5.00 net; Sheep or Half-Morocco, $6.00 net.

VOLUME I. CONTAINS:

Hygiene.—Fevers (Ephemeral, Simple Continued, Typhus, Typhoid, Epidemic Cerebrospinal Meningitis, and Relapsing)—Scarlatina, Measles, Rötheln, Variola, Varioloid, Vaccinia, Varicella, Mumps, Whooping cough, Anthrax, Hydrophobia, Trichinosis, Actinomycosis, Glanders, and Tetanus.—Tuberculosis, Scrofula, Syphilis, Diphtheria, Erysipelas, Malaria, Cholera, and Yellow Fever.—Nervous, Muscular, and Mental Diseases etc.

VOLUME II. CONTAINS:

Urine (Chemistry and Microscopy)—Kidney and Lungs.—Air-passages (Larynx and Bronchi) and Pleura.—Pharynx, (Esophagus, Stomach and Intestines (including Intestinal Parasites), Heart, Aorta, Arteries and Veins, —Peritoneum, Liver and Pancreas.—Diathetic Diseases (Rheumatism, Rheumatoid Arthritis, Gout, Lithæmia, and Diabetes.)—Blood and Spleen.—Inflammation, Embolism, Thrombosis, Fever, and Bacteriology.

The articles are not written as though addressed to students in lectures, but are exhaustive descriptions of diseases, with the newest facts as regards Causation, Symptomatology, Diagnosis, Prognosis, and Treatment, including a large number of approved formulæ. The recent advances made in the study of the bacterial origin of various diseases are fully described, as well as the bearing of the knowledge so gained upon prevention and cure. The subjects of Bacteriology as a whole and of Immunity are fully considered in a separate section.

Methods of diagnosis are given the most minute and careful attention, thus enabling the reader to learn the very latest methods of investigation without consulting works specially devoted to the subject.

CONTRIBUTORS.

Dr. J. S. Billings, Philadelphia.
Francis Delafield, New York.
Reginald H. Fitz, Boston.
James W. Holland, Philadelphia.
Henry M. Lyman, Chicago.
William Osler, Baltimore

Dr. William Pepper, Philadelphia.
W. Gilman Thompson, New York
W. H. Welch, Baltimore
James T. Whittaker, Cincinnati
James C. Wilson, Philadelphia.
Horatio C. Wood, Philadelphia.

" We reviewed the first volume of this work, and said: 'It is undoubtedly one of the best text-books on the practice of medicine which we possess.' A consideration of the second and last volume leads us to modify that verdict and to say that the completed work is, in our opinion, THE BEST of its kind it has ever been our fortune to see. It is complete, thorough, accurate, and clear. It is well written, well arranged, well printed, well illustrated, and well bound. It is a model of what the modern text-book should be."—*New York Medical Journal.*

" A library upon modern medical art. The work must promote the wider diffusion of sound knowledge."—*American Lancet.*

" A trusty counsellor for the practitioner or senior student, on which he may implicitly rely."—*Edinburgh Medical Journal.*

*AN AMERICAN TEXT-BOOK OF SURGERY. Edited by WILLIAM W. KEEN, M. D., LL.D., and J. WILLIAM WHITE, M. D., PH. D. Forming one handsome royal octavo volume of 1230 pages (10 x 7 inches), with 496 wood-cuts in text, and 37 colored and half-tone plates, many of them engraved from original photographs and drawings furnished by the authors. Price: Cloth, $7.00 net; Sheep or Half Morocco, $8.00 net.

THIRD EDITION, THOROUGHLY REVISED.

In the present edition, among the new topics introduced are a full consideration of serum-therapy; leucocytosis; post-operative insanity; the use of dry heat at high temperatures; Krönlein's method of locating the cerebral fissures; Hoffa's and Lorenz's operations of congenital dislocations of the hip; Allis's researches on dislocations of the hip-joint; lumbar puncture; the forcible reposition of the spine in Pott's disease; the treatment of exophthalmic goiter; the surgery of typhoid fever; gastrectomy and other operations on the stomach; new methods of operating upon the intestines; the use of Kelly's rectal specula; the surgery of the ureter; Schleich's infiltration-method and the use of eucain for local anesthesia; Krause's method of skin-grafting; the newer methods of disinfecting the hands; the use of gloves, etc. The sections on Appendicitis, on Fractures, and on Gynecological Operations have been revised and enlarged. A considerable number of new illustrations have been added, and enhance the value of the work.

The text of the entire book has been submitted to all the authors for their mutual criticism and revision—an idea in book-making that is entirely new and original. The book as a whole, therefore, expresses on all the important surgical topics of the day the consensus of opinion of the eminent surgeons who have joined in its preparation.

One of the most attractive features of the book is its illustrations. Very many of them are original and faithful reproductions of photographs taken directly from patients or from specimens.

CONTRIBUTORS:

Dr. Phineas S. Conner, Cincinnati.
Frederic S. Dennis, New York.
William W. Keen, Philadelphia.
Charles B. Nancrede, Ann Arbor, Mich.
Roswell Park, Buffalo, New York.
Lewis S. Pilcher, New York.
Dr. Nicholas Senn, Chicago.
Francis J. Shepherd, Montreal, Canada.
Lewis A. Stimson, New York.
J. Collins Warren, Boston.
J. William White, Philadelphia.

"If this text-book is a fair reflex of the present position of American surgery, we must admit it is of a very high order of merit, and that English surgeons will have to look very carefully to their laurels if they are to preserve a position in the van of surgical practice."—*London Lancet.*

Personally, I should not mind it being called THE TEXT-BOOK (instead of A TEXT-BOOK), for I know of no single volume which contains so readable and complete an account of the science and art of Surgery as this does."—EDMUND OWEN, F. R. C. S., *Member of the Board of Examiners of the Royal College of Surgeons, England*

* **AN AMERICAN TEXT-BOOK OF GYNECOLOGY, MEDICAL AND SURGICAL,** for the use of Students and Practitioners. Edited by J. M. BALDY, M. D. Forming a handsome royal-octavo volume of 718 pages, with 341 illustrations in the text and 38 colored and halftone plates. Prices: Cloth, $6.00 net; Sheep or Half-Morocco, $7.00 net.

SECOND EDITION, THOROUGHLY REVISED.

In this volume all anatomical descriptions, excepting those essential to a clear understanding of the text, have been omitted, the illustrations being largely depended upon to elucidate the anatomy of the parts. This work, which is thoroughly practical in its teachings, is intended, as its title implies, to be a working text-book for physicians and students. A clear line of treatment has been laid down in every case, and although no attempt has been made to discuss mooted points, still the most important of these have been noted and explained. The operations recommended are fully illustrated, so that the reader, having a picture of the procedure described in the text under his eye, cannot fail to grasp the idea. All extraneous matter and discussions have been carefully excluded, the attempt being made to allow no unnecessary details to cumber the text. The subject-matter is brought up to date at every point, and the work is as nearly as possible the combined opinions of the ten specialists who figure as the authors.

In the revised edition much new material has been added, and some of the old eliminated or modified. More than forty of the old illustrations have been replaced by new ones, which add very materially to the elucidation of the text, as they picture methods, not specimens. The chapters on technique and after-treatment have been considerably enlarged, and the portions devoted to plastic work have been so greatly improved as to be practically new. Hysterectomy has been rewritten, and all the descriptions of operative procedures have been carefully revised and fully illustrated.

CONTRIBUTORS:

Dr. Henry T. Byford.
John M. Baldy.
Edwin Cragin.
J. H. Etheridge.
William Goodell.

Dr. Howard A. Kelly.
Florian Krug.
E. E. Montgomery.
William R. Pryor.
George M. Tuttle.

"The most notable contribution to gynecological literature since 1887, ... and the most complete exponent of gynecology which we have. No subject seems to have been neglected, ... and the gynecologist and surgeon, and the general practitioner who has any desire to practise diseases of women, will find it of practical value. In the matter of illustrations and plates the book surpasses anything we have seen."—*Boston Medical and Surgical Journal.*

"A thoroughly modern text-book, and gives reliable and well-tempered advice and instruction."—*Edinburgh Medical Journal.*

"The harmony of its conclusions and the homogeneity of its style give it an individuality which suggests a single rather than a multiple authorship."—*Annals of Surgery.*

"It must command attention and respect as a worthy representation of our advanced clinical teaching."—*American Journal of Medical Sciences.*

*AN AMERICAN TEXT-BOOK OF THE DISEASES OF CHILDREN. By American Teachers. Edited by LOUIS STARR, M. D., assisted by THOMPSON S. WESTCOTT, M. D. In one handsome royal-8vo volume of 1244 pages, profusely illustrated with wood-cuts, half-tone and colored plates. Net Prices: Cloth, $7.00; Sheep or Half-Morocco, $8.00.

SECOND EDITION, REVISED AND ENLARGED.

The plan of this work embraces a series of original articles written by some sixty well-known pædiatrists, representing collectively the teachings of the most prominent medical schools and colleges of America. The work is intended to be a PRACTICAL book, suitable for constant and handy reference by the practitioner and the advanced student.

Especial attention has been given to the latest accepted teachings upon the etiology, symptoms, pathology, diagnosis, and treatment of the disorders of children, with the introduction of many special formulæ and therapeutic procedures.

In this new edition the whole subject matter has been carefully revised, new articles added, some original papers emended, and a number entirely rewritten. The new articles include "Modified Milk and Percentage Milk-Mixtures," "Lithemia," and a section on "Orthopedics." Those rewritten are "Typhoid Fever," "Rubella," "Chicken-pox," "Tuberculous Meningitis," "Hydrocephalus," and "Scurvy;" while extensive revision has been made in "Infant Feeding," "Measles," "Diphtheria," and "Cretinism." The volume has thus been much increased in size by the introduction of fresh material.

CONTRIBUTORS:

Dr. S. S. Adams, Washington.
John Ashhurst, Jr., Philadelphia.
A. D. Blackader, Montreal, Canada.
David Bovaird, New York.
Dillon Brown, New York.
Edward M. Buckingham, Boston.
Charles W. Burr, Philadelphia.
W. E. Casselberry, Chicago.
Henry Dwight Chapin, New York.
W. S. Christopher, Chicago.
Archibald Church, Chicago.
Floyd M. Crandall, New York.
Andrew F. Currier, New York.
Roland G. Curtin, Philadelphia
J. M. DaCosta, Philadelphia.
I. N. Danforth, Chicago.
Edward P. Davis, Philadelphia.
John B. Deaver, Philadelphia.
G. E. de Schweinitz, Philadelphia.
John Dorning, New York.
Charles Warrington Earle, Chicago.
Wm. A. Edwards, San Diego, Cal.
F. Forchheimer, Cincinnati.
J. Henry Fruitnight, New York.
J. P. Crozer Griffith, Philadelphia.
W. A. Hardaway. St. Louis.
M. P Hatfield, Chicago.
Barton Cooke Hirst, Philadelphia.
H. Illoway, Cincinnati.
Henry Jackson, Boston.
Charles G. Jennings, Detroit
Henry Koplik, New York.

Dr. Thomas S. Latimer, Baltimore.
Albert R. Leeds, Hoboken, N. J.
J. Hendrie Lloyd, Philadelphia.
George Roe Lockwood, New York.
Henry M. Lyman, Chicago.
Francis T. Miles, Baltimore.
Charles K Mills, Philadelphia.
James E. Moore, Minneapolis.
F. Gordon Morrill, Boston.
John H. Musser, Philadelphia.
Thomas R. Neilson, Philadelphia.
W. P. Northrup, New York.
William Osler, Baltimore.
Frederick A. Packard, Philadelphia.
William Pepper, Philadelphia.
Frederick Peterson, New York.
W. T. Plant, Syracuse, New York.
William M. Powell, Atlantic City.
B. K. Rachford, Cincinnati.
B. Alexander Randall, Philadelphia.
Edward O. Shakespeare, Philadelphia
F. C. Shattuck, Boston.
J. Lewis Smith, New York.
Louis Starr, Philadelphia.
M. Allen Starr, New York.
Charles W. Townsend, Boston.
James Tyson, Philadelphia.
W. S. Thayer, Baltimore.
Victor C. Vaughan, Ann Arbor, Mich
Thompson S. Westcott, Philadelphia.
Henry R. Wharton, Philadelphia.
J. William White, Philadelphia.
J. C. Wilson, Philadelphia.

***AN AMERICAN TEXT-BOOK OF GENITO-URINARY AND SKIN DISEASES.** By 47 Eminent Specialists and Teachers. Edited by L. BOLTON BANGS, M. D., Professor of Genito-Urinary Surgery, University and Bellevue Hospital Medical College, New York; and W. A. HARDAWAY, M. D., Professor of Diseases of the Skin, Missouri Medical College. Imperial octavo volume of 1229 pages, with 300 engravings and 20 full-page colored plates. Cloth, $7.00 net; Sheep or Half Morocco, $8.00 net.

This addition to the series of "American Text-Books," it is confidently believed, will meet the requirements of both students and practitioners, giving, as it does, a comprehensive and detailed presentation of the Diseases of the Genito-Urinary Organs, of the Venereal Diseases, and of the Affections of the Skin.

Having secured the collaboration of well-known authorities in the branches represented in the undertaking, the editors have not restricted the contributors in regard to the particular views set forth, but have offered every facility for the free expression of their individual opinions. The work will therefore be found to be original, yet homogeneous and fully representative of the several departments of medical science with which it is concerned.

CONTRIBUTORS:

Dr. Chas. W. Allen, New York.
I. E. Atkinson, Baltimore.
L. Bolton Bangs, New York.
P. R. Bolton, New York.
Lewis C. Bosher, Richmond, Va.
John T. Bowen, Boston.
J. Abbott Cantrell, Philadelphia.
William T. Corlett, Cleveland, Ohio.
B. Farquhar Curtis, New York.
Condict W. Cutler, New York.
Isadore Dyer, New Orleans.
Christian Fenger, Chicago.
John A. Fordyce, New York.
Eugene Fuller, New York.
R. H. Greene, New York.
Joseph Grindon, St. Louis.
Graeme M. Hammond, New York.
W. A. Hardaway, St. Louis.
M. B. Hartzell, Philadelphia.
Louis Heitzmann, New York.
James S. Howe, Boston.
George T. Jackson, New York.
Abraham Jacobi, New York.
James C. Johnston, New York.

Dr. Hermann G. Klotz, New York.
J. H. Linsley, Burlington, Vt.
G. F. Lydston, Chicago.
Hartwell N. Lyon, St. Louis.
Edward Martin, Philadelphia.
D. G. Montgomery, San Francisco.
James Pedersen, New York.
S. Pollitzer, New York.
Thomas R. Pooley, New York.
A. R. Robinson, New York.
A. E. Regensburger, San Francisco.
Francis J. Shepherd, Montreal, Can.
S. C. Stanton, Chicago, Ill.
Emmanuel J. Stout, Philadelphia.
Alonzo E. Taylor, Philadelphia.
Robert W. Taylor, New York.
Paul Thorndike, Boston.
H. Tuholske, St. Louis.
Arthur Van Harlingen, Philadelphia.
Francis S. Watson, Boston.
J. William White, Philadelphia.
J. McF. Winfield, Brooklyn.
Alfred C. Wood, Philadelphia.

"This voluminous work is thoroughly up to date, and the chapters on genito-urinary diseases are especially valuable. The illustrations are fine and are mostly original. The section on dermatology is concise and in every way admirable."—*Journal of the American Medical Association.*

"This volume is one of the best yet issued of the publisher's series of 'American Text-Books.' The list of contributors represents an extraordinary array of talent and extended experience. The book will easily take the place in comprehensiveness and value of the half dozen or more costly works on these subjects which have hitherto been necessary to a well-equipped library."—*New York Polyclinic.*

*AN AMERICAN TEXT-BOOK OF DISEASES OF THE EYE, EAR, NOSE, AND THROAT. Edited by GEORGE E. DE SCHWEINITZ, A. M., M. D., Professor of Ophthalmology, Jefferson Medical College; and B. ALEXANDER RANDALL, A. M., M. D., Clinical Professor of Diseases of the Ear, University of Pennsylvania. One handsome imperial octavo volume of 1251 pages; 766 illustrations, 59 of them colored. Prices: Cloth, $7.00 net; Sheep or Half-Morocco, $8.00 net.

Just Issued.

The present work is the only book ever published embracing diseases of the intimately related organs of the eye, ear, nose, and throat. Its special claim to favor is based on encyclopedic, authoritative, and practical treatment of the subjects.

Each section of the book has been entrusted to an author who is specially identified with the subject on which he writes, and who therefore presents his case in the manner of an expert. Uniformity is secured and overlapping prevented by careful editing and by a system of cross-references which forms a special feature of the volume, enabling the reader to come into touch with all that is said on any subject in different portions of the book.

Particular emphasis is laid on the most approved methods of treatment, so that the book shall be one to which the student and practitioner can refer for information in practical work. Anatomical and physiological problems, also, are fully discussed for the benefit of those who desire to investigate the more abstruse problems of the subject.

CONTRIBUTORS:

Dr. Henry A. Alderton, Brooklyn.
Harrison Allen, Philadelphia.
Frank Allport, Chicago.
Morris J. Asch, New York.
S. C. Ayres, Cincinnati.
R. O. Beard, Minneapolis.
Clarence J. Blake, Boston.
Arthur A. Bliss, Philadelphia.
Albert P. Brubaker, Philadelphia.
J. H. Bryan, Washington, D. C.
Albert H. Buck, New York.
F. Buller, Montreal, Can.
Swan M. Burnett, Washington, D C.
Flemming Carrow, Ann Arbor, Mich.
W. E. Casselberry, Chicago.
Colman W. Cutler, New York.
Edward B. Dench, New York.
William S. Dennett, New York.
George E. de Schweinitz, Philadelphia.
Alexander Duane, New York.
John W. Farlow, Boston, Mass.
Walter J. Freeman, Philadelphia.
H. Gifford, Omaha, Neb.
W. C. Glasgow, St. Louis.
J. Orne Green, Boston.
Ward A. Holden, New York.
Christian R. Holmes, Cincinnati.
William E. Hopkins, San Francisco.
F. C. Hotz, Chicago.
Lucien Howe, Buffalo, N. Y.

Dr. Alvin A. Hubbell, Buffalo, N. Y.
Edward Jackson, Philadelphia.
J. Ellis Jennings, St. Louis.
Herman Knapp, New York.
Chas. W. Kollock, Charleston, S. C.
G. A. Leland, Boston.
J. A. Lippincott, Pittsburg, Pa.
G. Hudson Makuen, Philadelphia.
John H. McCollom, Boston.
H. G. Miller, Providence, R. I.
B. L. Milliken, Cleveland, Ohio.
Robert C. Myles, New York.
James E. Newcomb, New York.
R. J. Phillips, Philadelphia.
George A. Piersol, Philadelphia.
W. P. Porcher, Charleston, S. C.
B. Alex. Randall, Philadelphia.
Robert L. Randolph, Baltimore.
John O. Roe, Rochester, N. Y
Charles E. de M. Sajous, Philadelphia.
J. E. Sheppard, Brooklyn, N. Y.
E. L. Shurly, Detroit, Mich.
William M. Sweet, Philadelphia.
Samuel Theobald, Baltimore, Md.
A. G. Thomson, Philadelphia.
Clarence A. Veasey, Philadelphia.
John E. Weeks, New York.
Casey A. Wood, Chicago, Ill.
Jonathan Wright, Brooklyn.
H. V. Würdemann, Milwaukee, Wis.

* **AN AMERICAN YEAR-BOOK OF MEDICINE AND SURGERY.** A Yearly Digest of Scientific Progress and Authoritative Opinion in all branches of Medicine and Surgery, drawn from journals, monographs, and text-books of the leading American and Foreign authors and investigators. Collected and arranged, with critical editorial comments, by eminent American specialists and teachers, under the general editorial charge of GEORGE M. GOULD, M. D. Volumes for 1896, '97, '98, and '99 each a handsome imperial octavo volume of about 1200 pages. Prices: Cloth, $6.50 net; Half-Morocco, $7.50 net. Year-Book for 1900 in two octavo volumes of about 600 pages each. Prices per volume: Cloth, $3.00 net; Half-Morocco, $3.75 net.

In Two Volumes. No Increase in Price.

In response to a widespread demand from the medical profession, the publisher of the "American Year-Book of Medicine and Surgery" has decided to issue that well-known work in two volumes, Vol. I. treating of **General Medicine**, Vol. II. of **General Surgery**. Each volume is complete in itself, and the work is sold either separately or in sets.

This division is made in such a way as to appeal to physicians from a class standpoint, one volume being distinctly medical, and the other distinctly surgical. This arrangement has a two-fold advantage. To the physician who uses the entire book, it offers an increased amount of matter in the most convenient form for easy consultation, and without any increase in price; while the man who wants either the medical or the surgical section alone secures the complete consideration of his branch without the necessity of purchasing matter for which he has no use.

CONTRIBUTORS:

Vol. I.
Dr. Samuel W. Abbott, Boston.
Archibald Church, Chicago.
Louis A. Duhring, Philadelphia.
D. L. Edsall, Philadelphia.
Alfred Hand, Jr., Philadelphia.
M. B. Hartzell, Philadelphia.
Reid Hunt, Baltimore.
Wyatt Johnston, Montreal.
Walter Jones, Baltimore.
David Riesman, Philadelphia.
Louis Starr, Philadelphia.
Alfred Stengel, Philadelphia.
A. A. Stevens, Philadelphia.
G. N. Stewart, Cleveland.
Reynold W. Wilcox, New York City.

Vol. II.
Dr. J. Montgomery Baldy, Philadelphia.
Charles H. Burnett, Philadelphia.
J. Chalmers DaCosta, Philadelphia.
W. A. N. Dorland, Philadelphia.
Virgil P. Gibney, New York City.
C. H. Hamann, Cleveland.
Howard F. Hansell, Philadelphia.
Barton Cooke Hirst, Philadelphia.
E. Fletcher Ingals, Chicago.
W. W. Keen, Philadelphia.
Henry G. Ohls, Chicago.
Wendell Reber, Philadelphia.
J. Hilton Waterman, New York City.

"It is difficult to know which to admire most—the research and industry of the distinguished band of experts whom Dr. Gould has enlisted in the service of the Year-Book, or the wealth and abundance of the contributions to every department of science that have been deemed worthy of analysis. . . . It is much more than a mere compilation of abstracts, for, as each section is entrusted to experienced and able contributors, the reader has the advantage of certain critical commentaries and expositions . . . proceeding from writers fully qualified to perform these tasks. . . . It is emphatically a book which should find a place in every medical library, and is in several respects more useful than the famous 'Jahrbucher' of Germany."—*London Lancet.*

* ANOMALIES AND CURIOSITIES OF MEDICINE. By GEORGE M. GOULD, M.D., and WALTER L. PYLE, M.D. An encyclopedic collection of are and extraordinary cases and of the most striking instances of abnormality in all branches of Medicine and Surgery, derived from an exhaustive research of medical literature from its origin to the present day, abstracted, classified, annotated, and indexed. Handsome imperial octavo volume of 968 pages, with 295 engravings in the text, and 12 full-page plates. Cloth, $3.00 net; Half-Morocco, $4.00 net.

POPULAR EDITION REDUCED FROM $6.00 to $3.00.

In view of the great success of this magnificent work, the publisher has decided to issue a "Popular Edition" at a price so low that it may be procured by every student and practitioner of medicine. Notwithstanding the great reduction in price, there will be no depreciation in the excellence of typography, paper, and binding that characterized the earlier editions.

Several years of exhaustive research have been spent by the authors in the great medical libraries of the United States and Europe in collecting the material for this work. **Medical literature of all ages and all languages has** been carefully searched, as a glance at the Bibliographic Index will show. The facts, which will be of **extreme value to the author and lecturer,** have been arranged and annotated, and full reference footnotes given.

"One of the most valuable contributions ever made to medical literature. It is, so far as we know, absolutely unique, and every page is as fascinating as a novel. Not alone for the medical profession has this volume value: it will serve as a book of reference for all who are interested in general scientific, sociologic, or medico-legal topics."—*Brooklyn Medical Journal.*

NERVOUS AND MENTAL DISEASES. By ARCHIBALD CHURCH, M. D., Professor of Clinical Neurology, Mental Diseases, and Medical Jurisprudence, Northwestern University Medical School; and FREDERICK PETERSON, M.D., Clinical Professor of Mental Diseases, Woman's Medical College, New York. Handsome octavo volume of 843 pages, with over 300 illustrations. Prices: Cloth, $5.00 net; Half-Morocco, $6.00 net.

Second Edition.

This book is intended to furnish students and practitioners with a practical, working knowledge of nervous and mental diseases. Written by men of wide experience and authority, it presents the many recent additions to the subject. The book is not filled with an extended dissertation on anatomy and pathology, but, treating these points in connection with special conditions, it lays particular stress on methods of examination, diagnosis, and treatment. In this respect the work is unusually complete and valuable, laying down the definite courses of procedure which the authors have found to be most generally satisfactory.

"The work is an epitome of what is to-day known of nervous diseases prepared for the student and practitioner in the light of the author's experience . . . We believe that no work presents the difficult subject of insanity in such a reasonable and readable way."—*Chicago Medical Recorder.*

DISEASES OF THE NOSE AND THROAT. By D. BRADEN KYLE, M. D., Clinical Professor of Laryngology and Rhinology, Jefferson Medical College, Philadelphia; Consulting Laryngologist, Rhinologist, and Otologist, St. Agnes' Hospital. Octavo volume of 646 pages, with over 150 illustrations and 6 lithographic plates. Cloth, $4.00 net; Half-Morocco, $5.00 net.

Just Issued.

This book presents the subject of Diseases of the Nose and Throat in as concise a manner as is consistent with clearness, keeping in mind the needs of the student and general practitioner as well as those of the specialist. The arrangement and classification are based on modern pathology, and the pathological views advanced are supported by drawings of microscopical sections made in the author's own laboratory. These and the other illustrations are particularly fine, being chiefly original. With the practical purpose of the book in mind, extended consideration has been given to details of treatment, each disease being considered in full, and definite courses being laid down to meet special conditions and symptoms.

" It is a thorough, full, and systematic treatise, so classified and arranged as greatly to facilitate the teaching of laryngology and rhinology to classes, and must prove most convenient and satisfactory as a reference book, both for students and practitioners."—*International Medical Magazine.*

THE HYGIENE OF TRANSMISSIBLE DISEASES; their Causation, Modes of Dissemination, and Methods of Prevention. By A. C. ABBOTT, M. D., Professor of Hygiene in the University of Pennsylvania; Director of the Laboratory of Hygiene. Octavo volume of 311 pages, with charts and maps, and numerous illustrations. Cloth, $2.00 net.

Just Issued.

It is not the purpose of this work to present the subject of Hygiene in the comprehensive sense ordinarily implied by the word, but rather to deal directly with but a section, certainly not the least important, of the subject—viz., that embracing a knowledge of the preventable specific diseases. The book aims to furnish information concerning the detailed management of transmissible diseases. Incidentally there are discussed those numerous and varied factors that have not only a direct bearing upon the incidence and suppression of such diseases, but are of general sanitary importance as well.

" The work is admirable in conception and no less so in execution. It is a practical work, simply and lucidly written, and it should prove a most helpful aid in that department of medicine which is becoming daily of increasing importance and application—namely, prophylaxis."—*Philadelphia Medical Journal.*

" It is scientific, but not too technical; it is as complete as our present-day knowledge of hygiene and sanitation allows, and it is in harmony with the efforts of the profession, which are tending more and more to methods of prophylaxis. For the student and for the practitioner it is well nigh indispensable."—*Medical News,* New York.

A TEXT-BOOK OF EMBRYOLOGY. By JOHN C. HEISLER, M. D., Professor of Anatomy in the Medico-Chirurgical College, Philadelphia Octavo volume of 405 pages, with 190 illustrations, 26 in colors. Cloth $2.50 net.

Just Issued.

The facts of embryology having acquired in recent years such great interest in connection with the teaching and with the proper comprehension of human anatomy, it is of first importance to the student of medicine that a concise and yet sufficiently full text-book upon the subject be available. It was with the aim of presenting such a book that this volume was written, the author, in his experience as a teacher of anatomy, having been impressed with the fact that students were seriously handicapped in their study of the subject of embryology by the lack of a text-book full enough to be intelligible, and yet without that minuteness of detail which characterizes the larger treatises, and which so often serves only to confuse and discourage the beginner.

"In short, the book is written to fill a want which has distinctly existed and which it definitely meets; commendation greater than this it is not possible to give to anything."— *Medical News*, New York.

A MANUAL OF DISEASES OF THE EYE. By EDWARD JACKSON, A. M., M. D., sometime Professor of Diseases of the Eye in the Philadelphia Polyclinic and College for Graduates in Medicine. 12mo, 604 pages, with 178 illustrations from drawings by the author. Cloth, $2.50 net.

Just Issued.

This book is intended to meet the needs of the general practitioner of medicine and the beginner in ophthalmology. More attention is given to the conditions that must be met and dealt with early in ophthalmic practice than to the rarer diseases and more difficult operations that may come later.

It is designed to furnish efficient aid in the actual work of dealing with disease, and therefore gives the place of first importance to the recognition and management of the conditions that present themselves in actual clinical work.

LECTURES ON THE PRINCIPLES OF SURGERY. By CHARLES B. NANCREDE, M. D., LL.D., Professor of Surgery and of Clinical Surgery, University of Michigan, Ann Arbor. Handsome octavo, 398 pages, illustrated. Cloth, $2.50 net.

Just Issued.

The present book is based on the lectures delivered by Dr. Nancrede to his undergraduate classes, and is intended as a text-book for students and a practical help for teachers. By the careful elimination of unnecessary details of pathology, bacteriology, etc., which are amply provided for in other courses of study, space is gained for a more extended consideration of the Principles of Surgery in themselves, and of the application of these principles to methods of practice.

A TEXT-BOOK OF PATHOLOGY. By ALFRED STENGEL, M. D., Professor of Clinical Medicine in the University of Pennsylvania; Physician to the Philadelphia Hospital; Physician to the Children's Hospital, Philadelphia. Handsome octavo volume of 848 pages, with 362 illustrations, many of which are in colors. Prices: Cloth, $4.00 net; Half-Morocco, $5.00 net.

Second Edition.

In this work the practical application of pathological facts to clinical medicine is considered more fully than is customary in works on pathology. While the subject of pathology is treated in the broadest way consistent with the size of the book, an effort has been made to present the subject from the point of view of the clinician. The general relations of bacteriology to pathology are discussed at considerable length, as the importance of these branches deserves. It will be found that the recent knowledge is fully considered, as well as older and more widely-known facts.

"I consider the work abreast of modern pathology, and useful to both students and practitioners. It presents in a concise and well-considered form the essential facts of general and special pathological anatomy, with more than usual emphasis upon pathological physiology."
—WILLIAM H. WELCH, *Professor of Pathology, Johns Hopkins University, Baltimore, Md.*

"I regard it as the most serviceable text-book for students on this subject yet written by an American author."—L. HEKTOEN, *Professor of Pathology, Rush Medical College, Chicago, Ill.*

A TEXT-BOOK OF OBSTETRICS. By BARTON COOKE HIRST, M.D., Professor of Obstetrics in the University of Pennsylvania. Handsome octavo volume of 846 pages, with 618 illustrations and seven colored plates. Prices: Cloth, $5.00 net; Half-Morocco, $6.00 net.

Second Edition.

This work, which has been in course of preparation for several years, is intended as an ideal text-book for the student no less than an advanced treatise for the obstetrician and for general practitioners. It represents the very latest teaching in the practice of obstetrics by a man of extended experience and recognized authority. The book emphasizes especially, as a work on obstetrics should, the practical side of the subject, and to this end presents an unusually large collection of illustrations. A great number of these are new and original, and the whole collection will form a complete atlas of obstetrical practice. An extremely valuable feature of the book is the large number of references to cases, authorities, sources, etc., forming, as it does, a valuable bibliography of the most recent and authoritative literature on the subject of obstetrics. As already stated, this work records the wide practical experience of the author, which fact, combined with the brilliant presentation of the subject, will doubtless render this one of the most notable books on obstetrics that has yet appeared.

"The illustrations are numerous and are works of art, many of them appearing for the first time. The arrangement of the subject-matter, the foot-notes, and index are beyond criticism. The author's style, though condensed, is singularly clear, so that it is never necessary to re-read a sentence in order to grasp its meaning. As a true model of what a modern text-book in obstetrics should be, we feel justified in affirming that Dr. Hirst's book is without a rival."—*New York Medical Record.*

A TEXT-BOOK OF THE PRACTICE OF MEDICINE. By JAMES M. ANDERS, M. D., PH.D., LL.D., Professor of the Practice of Medicine and of Clinical Medicine, Medico-Chirurgical College, Philadelphia. In one handsome octavo volume of 1292 pages, fully illustrated. Cloth, $5.50 net; Sheep or Half-Morocco, $6.50 net.

THIRD EDITION, THOROUGHLY REVISED.

The present edition is the result of a careful and thorough revision. A few new subjects have been introduced : Glandular Fever, Ether-pneumonia, Splenic Anemia, Meralgia Paresthetica, and Periodic Paralysis. The affections that have been substantially rewritten are: Plague, Malta Fever, Diseases of the Thymus Gland, Liver Cirrhoses, and Progressive Spinal Muscular Atrophy. The following articles have been extensively revised: Typhoid Fever, Yellow Fever, Lobar Pneumonia, Dengue, Tuberculosis, Diabetes Mellitus, Gout, Arthritis Deformans, Autumnal Catarrh, Diseases of the Circulatory System, more particularly Hypertrophy and Dilatation of the Heart, Arteriosclerosis and Thoracic Aneurysm, Pancreatic Hemorrhage, Jaundice, Acute Peritonitis, Acute Yellow Atrophy, Hematoma of Dura Mater, and Scleroses of the Brain. The preliminary chapter on Nervous Diseases is new, and deals with the subject of localization and the various methods of investigating nervous affections.

"It is an excellent book—concise, comprehensive, thorough, and up to date. It is a credit to you; but, more than that, it is a credit to the profession of Philadelphia—to us."
—JAMES C. WILSON, *Professor of the Practice of Medicine and Clinical Medicine, Jefferson Medical College, Philadelphia.*

"The book can be unreservedly recommended to students and practitioners as a safe, full compendium of the knowledge of internal medicine of the present day . . . It is a work thoroughly modern in every sense."—*Medical News*, New York.

DISEASES OF THE STOMACH. By WILLIAM W. VAN VALZAH, M. D., Professor of General Medicine and Diseases of the Digestive System and the Blood, New York Polyclinic; and J. DOUGLAS NISBET, M. D., Adjunct Professor of General Medicine and Diseases of the Digestive System and the Blood, New York Polyclinic. Octavo volume of 674 pages, illustrated. Cloth, $3.50 net.

An eminently practical book, intended as a guide to the student, an aid to the physician, and a contribution to scientific medicine. It aims to give a complete description of the modern methods of diagnosis and treatment of diseases of the stomach, and to reconstruct the pathology of the stomach in keeping with the revelations of scientific research. The book is clear, practical, and complete, and contains the results of the authors' investigations and of their extensive experience as specialists. Particular attention is given to the important subject of dietetic treatment. The diet-lists are very complete, and are so arranged that selections can readily be made to suit individual cases.

"This is the most satisfactory work on the subject in the English language."—*Chicago Medical Recorder.*

"The article on diet and general medication is one of the most valuable in the book, and should be read by every practising physician."—*New York Medical Journal.*

SURGICAL DIAGNOSIS AND TREATMENT. By J. W. MAC-DONALD, M. D., Edin., F. R. C. S., Edin., Professor of the Practice of Surgery and of Clinical Surgery in Hamline University; Visiting Surgeon to St. Barnabas' Hospital, Minneapolis, etc. Handsome octavo volume of 800 pages, profusely illustrated. Cloth, $5.00 net; Half-Morocco, $6.00 net.

This work aims in a comprehensive manner to furnish a guide in matters of surgical diagnosis. It sets forth in a systematic way the necessities of examinations and the proper methods of making them. The various portions of the body are then taken up in order and the diseases and injuries thereof succinctly considered and the treatment briefly indicated. Practically all the modern and approved operations are described with thoroughness and clearness. The work concludes with a chapter on the use of the Röntgen rays in surgery.

"The work is brimful of just the kind of practical information that is useful alike to students and practitioners. It is a pleasure to commend the book because of its intrinsic value to the medical practitioner."—*Cincinnati Lancet-Clinic.*

PATHOLOGICAL TECHNIQUE. A Practical Manual for Laboratory Work in Pathology, Bacteriology, and Morbid Anatomy, with chapters on Post-Mortem Technique and the Performance of Autopsies. By FRANK B. MALLORY, A. M., M. D., Assistant Professor of Pathology, Harvard University Medical School, Boston; and JAMES H. WRIGHT, A. M., M. D., Instructor in Pathology, Harvard University Medical School, Boston. Octavo volume of 396 pages, handsomely illustrated. Cloth, $2.50 net.

This book is designed especially for practical use in pathological laboratories, both as a guide to beginners and as a source of reference for the advanced. The book will also meet the wants of practitioners who have opportunity to do general pathological work. Besides the methods of post-mortem examinations and of bacteriological and histological investigations connected with autopsies, the special methods employed in clinical bacteriology and pathology have been fully discussed.

"One of the most complete works on the subject, and one which should be in the library of every physician who hopes to keep pace with the great advances made in pathology."—*Journal of American Medical Association.*

THE SURGICAL COMPLICATIONS AND SEQUELS OF TYPHOID FEVER. By WM. W. KEEN, M. D., LL.D., Professor of the Principles of Surgery and of Clinical Surgery, Jefferson Medical College, Philadelphia. Octavo volume of 386 pages, illustrated. Cloth, $3.00 net.

This monograph is the only one in any language covering the entire subject of the Surgical Complications and Sequels of Typhoid Fever. The work will prove to be of importance and interest not only to the general surgeon and physician, but also to many specialists—laryngologists, ophthalmologists, gynecologists, pathologists, and bacteriologists—as the subject has an important bearing upon each one of their spheres. The author's conclusions are based on reports of over 1700 cases, including practically all those recorded in the last fifty years. Reports of cases have been brought down to date, many having been added while the work was in press.

"This is probably the first and only work in the English language that gives the reader a clear view of what typhoid fever really is, and what it does and can do to the human organism. This book should be in the possession of every medical man in America."—*American Medico-Surgical Bulletin.*

MODERN SURGERY, GENERAL AND OPERATIVE. By JOHN CHALMERS DACOSTA, M.D., Professor of Practice of Surgery and Clinical Surgery, Jefferson Medical College, Philadelphia; Surgeon to the Philadelphia Hospital, etc. Handsome octavo volume of 911 pages, profusely illustrated. Cloth, $4.00 net; Half-Morocco, $5.00 net.

Second Edition, Rewritten and Greatly Enlarged.

The remarkable success attending DaCosta's Manual of Surgery, and the general favor with which it has been received, have led the author in this revision to produce a complete treatise on modern surgery along the same lines that made the former edition so successful. The book has been entirely rewritten and very much enlarged. The old edition has long been a favorite not only with students and teachers, but also with practising physicians and surgeons, and it is believed that the present work will find an even wider field of usefulness.

"We know of no small work on surgery in the English language which so well fulfils the requirements of the modern student."—*Medico-Chirurgical Journal,* Bristol, England.

"The author has presented concisely and accurately the principles of modern surgery. The book is a valuable one which can be recommended to students and is of great value to the general practitioner."—*American Journal of the Medical Sciences.*

A MANUAL OF ORTHOPEDIC SURGERY. By JAMES E. MOORE, M.D., Professor of Orthopedics and Adjunct Professor of Clinical Surgery, University of Minnesota, College of Medicine and Surgery. Octavo volume of 356 pages, with 177 beautiful illustrations from photographs made specially for this work. Cloth, $2.50 net.

A practical book based upon the author's experience, in which special stress is laid upon early diagnosis and treatment such as can be carried out by the general practitioner. The teachings of the author are in accordance with his belief that true conservatism is to be found in the middle course between the surgeon who operates too frequently and the orthopedist who seldom operates.

"A very demonstrative work, every illustration of which conveys a lesson. The work is a most excellent and commendable one, which we can certainly endorse with pleasure."— *St. Louis Medical and Surgical Journal.*

ELEMENTARY BANDAGING AND SURGICAL DRESSING. With Directions concerning the Immediate Treatment of Cases of Emergency. For the use of Dressers and Nurses. By WALTER PYE, F.R.C.S., late Surgeon to St. Mary's Hospital, London. Small 12mo, with over 80 illustrations. Cloth, flexible covers, 75 cents net.

This little book is chiefly a condensation of those portions of Pye's "Surgical Handicraft" which deal with bandaging, splinting, etc., and of those which treat of the management in the first instance of cases of emergency. The directions given are thoroughly practical, and the book will prove extremely useful to students, surgical nurses, and dressers.

"The author writes well, the diagrams are clear, and the book itself is small and portable, although the paper and type are good."—*British Medical Journal.*

A TEXT-BOOK OF MATERIA MEDICA, THERAPEUTICS AND PHARMACOLOGY. By GEORGE F. BUTLER, PH.G., M.D.,

Professor of Materia Medica and of Clinical Medicine in the College of Physicians and Surgeons, Chicago; Professor of Materia Medica and Therapeutics, Northwestern University, Woman's Medical School, etc. Octavo, 874 pages, illustrated. Cloth, $4.00 net; Sheep, $5.00 net.

Third Edition, Thoroughly Revised.

A clear, concise, and practical text-book, adapted for permanent reference no less than for the requirements of the class-room.

The recent important additions made to our knowledge of the physiological action of drugs are fully discussed in the present edition. The book has been thoroughly revised and many additions have been made.

"Taken as a whole, the book may fairly be considered as one of the most satisfactory of any single-volume works on materia medica in the market."—*Journal of the American Medical Association.*

TUBERCULOSIS OF THE GENITO-URINARY ORGANS, MALE AND FEMALE. By NICHOLAS SENN, M.D., PH.D., LL.D.,

Professor of the Practice of Surgery and of Clinical Surgery, Rush Medical College, Chicago. Handsome octavo volume of 320 pages, illustrated. Cloth, $3.00 net.

Tuberculosis of the male and female genito-urinary organs is such a frequent, distressing, and fatal affection that a special treatise on the subject appears to fill a gap in medical literature. In the present work the bacteriology of the subject has received due attention, the modern resources employed in the differential diagnosis between tubercular and other inflammatory affections are fully described, and the medical and surgical therapeutics are discussed in detail.

"An important book upon an important subject, and written by a man of mature judgment and wide experience. The author has given us an instructive book upon one of the most important subjects of the day."—*Clinical Reporter.*

"A work which adds another to the many obligations the profession owes the talented author."—*Chicago Medical Recorder.*

A TEXT-BOOK OF DISEASES OF WOMEN. By CHARLES B. PENROSE, M.D., PH.D.,

Professor of Gynecology in the University of Pennsylvania; Surgeon to the Gynecean Hospital, Philadelphia. Octavo volume of 531 pages, with 217 illustrations, nearly all from drawings made for this work. Cloth, $3.75 net.

Third Edition, Revised.

In this work, which has been written for both the student of gynecology and the general practitioner, the author presents the best teaching of modern gynecology untrammelled by antiquated theories or methods of treatment. In most instances but one plan of treatment is recommended, to avoid confusing the student or the physician who consults the book for practical guidance.

"I shall value very highly the copy of Penrose's 'Diseases of Women' received. I have already recommended it to my class as THE BEST book."—HOWARD A. KELLY, *Professor of Gynecology and Obstetrics, Johns Hopkins University, Baltimore, Md.*

"The book is to be commended without reserve, not only to the student but to the general practitioner who wishes to have the latest and best modes of treatment explained with absolute clearness."—*Therapeutic Gazette.*

CATALOGUE OF MEDICAL WORKS. 25

SURGICAL PATHOLOGY AND THERAPEUTICS. By JOHN COLLINS WARREN, M. D., LL.D., Professor of Surgery, Medical Department Harvard University. Handsome octavo, 832 pages, with 136 relief and lithographic illustrations, 33 of which are printed in colors.

Second Edition,

with an Appendix devoted to the Scientific Aids to Surgical Diagnosis, and a series of articles on Regional Bacteriology. Cloth, $5.00 net; Half-Morocco, $6.00 net.

Without Exception, the Illustrations are the Best ever Seen in a Work of this Kind.

"A most striking and very excellent feature of this book is its illustrations. Without exception, from the point of accuracy and artistic merit, they are the best ever seen in a work of this kind. * * * Many of those representing microscopic pictures are so perfect in their coloring and detail as almost to give the beholder the impression that he is looking down the barrel of a microscope at a well-mounted section."—*Annals of Surgery*, Philadelphia.

"It is the handsomest specimen of book-making * * * that has ever been issued from the American medical press."—*American Journal of the Medical Sciences*, Philadelphia.

PATHOLOGY AND SURGICAL TREATMENT OF TUMORS. By N. SENN, M. D., Ph. D., LL. D., Professor of Practice of Surgery and of Clinical Surgery, Rush Medical College; Professor of Surgery, Chicago Polyclinic; Attending Surgeon to Presbyterian Hospital; Surgeon-in-Chief, St. Joseph's Hospital, Chicago. One volume of 710 pages, with 515 engravings, including full-page colored plates. New and enlarged Edition in Preparation.

Books specially devoted to this subject are few, and in our text-books and systems of surgery this part of surgical pathology is usually condensed to a degree incompatible with its scientific and clinical importance. The author spent many years in collecting the material for this work, and has taken great pains to present it in a manner that should prove useful as a text-book for the student, a work of reference for the practitioner, and a reliable guide for the surgeon.

"The most exhaustive of any recent book in English on this subject. It is well illustrated, and will doubtless remain as the principal monograph on the subject in our language for some years. The book is handsomely illustrated and printed, ... and the author has given a notable and lasting contribution to surgery."—*Journal of the American Medical Association*, Chicago.

LECTURES ON RENAL AND URINARY DISEASES. By ROBERT SAUNDBY, M. D., Edin., Fellow of the Royal College of Physicians, London, and of the Royal Medico-Chirurgical Society; Physician to the General Hospital. Octavo volume of 434 pages, with numerous illustrations and 4 colored plates. Cloth, $2.50 net.

"The volume makes a favorable impression at once. The style is clear and succinct. We cannot find any part of the subject in which the views expressed are not carefully thought out and fortified by evidence drawn from the most recent sources. The book may be cordially recommended."—*British Medical Journal*.

A HANDBOOK FOR NURSES. By J. K. WATSON, M. D., Edin., Assistant House-Surgeon, Sheffield Royal Hospital. American Edition, under the supervision of A. A. STEVENS, A. M., M. D., Professor of Pathology, Woman's Medical College, Philadelphia. 12mo, 413 pages, 73 illustrations. Cloth, $1.50 net.

This work aims to supply in one volume that information which so many nurses at the present time are trying to extract from various medical works, and to present that information in a suitable form. Nurses must necessarily acquire a certain amount of medical knowledge, and the author of this book has aimed judiciously to cater to this need with the object of directing the nurses' pursuit of medical information in proper and legitimate channels. The book represents an entirely new departure in nursing literature, insomuch as it contains useful information on medical and surgical matters hitherto only to be obtained from expensive works written expressly for medical men.

A NEW PRONOUNCING DICTIONARY OF MEDICINE, with Phonetic Pronunciation, Accentuation, Etymology, etc. By JOHN M. KEATING, M. D., LL.D., Fellow of the College of Physicians of Philadelphia; Editor "Cyclopædia of the Diseases of Children," etc.; and HENRY HAMILTON, with the Collaboration of J. CHALMERS DACOSTA, M. D., and FREDERICK A. PACKARD, M. D. One very attractive volume of over 800 pages. Second Revised Edition. Prices: Cloth, $5.00 net; Sheep or Half-Morocco, $6.00 net; with Denison's Patent Ready-Reference Index; without patent index, Cloth, $4.00 net; Sheep or Half-Morocco, $5.00 net.

PROFESSIONAL OPINIONS.

"I am much pleased with Keating's Dictionary, and shall take pleasure in recommending it to my classes."
HENRY M. LYMAN, M. D.,
Professor of Principles and Practice of Medicine, Rush Medical College, Chicago, Ill.

"I am convinced that it will be a very valuable adjunct to my study-table, convenient in size and sufficiently full for ordinary use."
C. A. LINDSLEY, M. D.,
Professor of Theory and Practice of Medicine, Medical Dept. Yale University; Secretary Connecticut State Board of Health, New Haven, Conn.

AUTOBIOGRAPHY OF SAMUEL D. GROSS, M. D., Emeritus Professor of Surgery in the Jefferson Medical College of Philadelphia, with Reminiscences of His Times and Contemporaries. Edited by his sons, SAMUEL W. GROSS, M. D., LL.D., and A. HALLER GROSS, A. M., of the Philadelphia Bar. Preceded by a Memoir of Dr. Gross, by the late Austin Flint, M. D., LL.D. In two handsome volumes, each containing over 400 pages, demy 8vo, extra cloth, gilt tops, with fine Frontispiece engraved on steel. Price per Volume, $2.50 net.

PRACTICAL POINTS IN NURSING. For Nurses in Private Practice. By EMILY A. M. STONEY, Graduate of the Training-School for Nurses, Lawrence, Mass.; Superintendent of the Training-School for Nurses, Carney Hospital, South Boston, Mass. 456 pages, handsomely illustrated with 73 engravings in the text, and 9 colored and half-tone plates. Cloth. Price, $1.75 net.

SECOND EDITION, THOROUGHLY REVISED.

In this volume the author explains, in popular language and in the shortest possible form, the entire range of *private* nursing as distinguished from *hospital* nursing, and the nurse is instructed how best to meet the various emergencies of medical and surgical cases when distant from medical or surgical aid or when thrown on her own resources.

An especially valuable feature of the work will be found in the directions to the nurse how to *improvise* everything ordinarily needed in the sick-room, where the embarrassment of the nurse, owing to the want of proper appliances, is frequently extreme.

The work has been logically divided into the following sections:

I. The Nurse: her responsibilities, qualifications, equipment, etc.
II. The Sick-Room: its selection, preparation, and management.
III. The Patient: duties of the nurse in medical, surgical, obstetric, and gynecologic cases.
IV. Nursing in Accidents and Emergencies.
V. Nursing in Special Medical Cases.
VI. Nursing of the New-born and Sick Children.
VII. Physiology and Descriptive Anatomy.

The APPENDIX contains much information in compact form that will be found of great value to the nurse, including Rules for Feeding the Sick; Recipes for Invalid Foods and Beverages; Tables of Weights and Measures; Table for Computing the Date of Labor; List of Abbreviations; Dose-List; and a full and complete Glossary of Medical Terms and Nursing Treatment.

"This is a well-written, eminently practical volume, which covers the entire range of private nursing as distinguished from hospital nursing, and instructs the nurse how best to meet the various emergencies which may arise and how to prepare everything ordinarily needed in the illness of her patient."—*American Journal of Obstetrics and Diseases of Women and Children*, Aug., 1896.

A TEXT-BOOK OF BACTERIOLOGY, including the Etiology and Prevention of Infective Diseases and an account of Yeasts and Moulds, Hæmatozoa, and Psorosperms. By EDGAR M. CROOKSHANK, M. B., Professor of Comparative Pathology and Bacteriology, King's College, London. A handsome octavo volume of 700 pages, with 273 engravings in the text, and 22 original and colored plates. Price, $6.50 net.

This book, though nominally a Fourth Edition of Professor Crookshank's "MANUAL OF BACTERIOLOGY," is practically a new work, the old one having been reconstructed, greatly enlarged, revised throughout, and largely rewritten, forming a text-book for the Bacteriological Laboratory, for Medical Officers of Health, and for Veterinary Inspectors.

MEDICAL DIAGNOSIS. By Dr. OSWALD VIERORDT, Professor of Medicine at the University of Heidelberg. Translated, with additions, from the Fifth Enlarged German Edition, with the author's permission, by FRANCIS H. STUART, A. M., M. D. In one handsome royal-octavo volume of 600 pages. 194 fine wood-cuts in the text, many of them in colors. Prices: Cloth, $4.00 net; Sheep or Half-Morocco, $5.00 net.

FOURTH AMERICAN EDITION, FROM THE FIFTH REVISED AND ENLARGED GERMAN EDITION.

In this work, as in no other hitherto published, are given full and accurate explanations of the phenomena observed at the bedside. It is distinctly a clinical work by a master teacher, characterized by thoroughness, fulness, and accuracy. It is a mine of information upon the points that are so often passed over without explanation. Especial attention has been given to the germ-theory as a factor in the origin of disease.

The present edition of this highly successful work has been translated from the fifth German edition. Many alterations have been made throughout the book, but especially in the sections on Gastric Digestion and the Nervous System.

It will be found that all the qualities which served to make the earlier editions so acceptable have been developed with the evolution of the work to its present form.

THE PICTORIAL ATLAS OF SKIN DISEASES AND SYPHILITIC AFFECTIONS. (American Edition.) Translation from the French. Edited by J. J. PRINGLE, M. B., F. R. C. P., Assistant Physician to, and Physician to the department for Diseases of the Skin at, the Middlesex Hospital, London. Photo-lithochromes from the famous models of dermatological and syphilitic cases in the Museum of the Saint-Louis Hospital, Paris, with explanatory wood-cuts and letter-press. In 12 Parts, at $3.00 per Part.

"Of all the atlases of skin diseases which have been published in recent years, the present one promises to be of greatest interest and value, especially from the standpoint of the general practitioner."—*American Medico-Surgical Bulletin*, Feb. 22, 1896.

"The introduction of explanatory wood-cuts in the text is a novel and most important feature which greatly furthers the easier understanding of the excellent plates, than which nothing, we venture to say, has been seen better in point of correctness, beauty, and general merit."—*New York Medical Journal*, Feb. 15, 1896.

"An interesting feature of the Atlas is the descriptive text, which is written for each picture by the physician who treated the case or at whose instigation the models have been made. We predict for this truly beautiful work a large circulation in all parts of the medical world where the names *St. Louis* and *Baretta* have preceded it."—*Medical Record*, N. Y., Feb. 1, 1896.

A TEXT-BOOK OF MECHANO-THERAPY (MASSAGE AND MEDICAL GYMNASTICS). By AXEL V. GRAFSTROM, B. Sc., M. D., late Lieutenant in the Royal Swedish Army; late House Physician, City Hospital, Blackwell's Island, New York. 12mo, 139 pages, illustrated. Cloth, $1.00 net.

DISEASES OF THE EYE. A Hand-Book of Ophthalmic Practice. By G. E. DE SCHWEINITZ, M. D., Professor of Ophthalmology in the Jefferson Medical College, Philadelphia, etc. A handsome royal-octavo volume of 696 pages, with 255 fine illustrations, many of which are original, and 2 chromo-lithographic plates. Prices: Cloth, $4.00 net; Sheep or Half-Morocco, $5.00 net.

THIRD EDITION, THOROUGHLY REVISED.

In the third edition of this text-book, destined, it is hoped, to meet the favorable reception which has been accorded to its predecessors, the work has been revised thoroughly, and much new matter has been introduced. Particular attention has been given to the important relations which micro-organisms bear to many ocular diseases. A number of special paragraphs on new subjects have been introduced, and certain articles, including a portion of the chapter on Operations, have been largely rewritten, or at least materially changed. A number of new illustrations have been added. The Appendix contains a full description of the method of determining the corneal astigmatism with the ophthalmometer of Javal and Schiötz, and the rotation of the eyes with the tropometer of Stevens.

"A work that will meet the requirements not only of the specialist, but of the general practitioner in a rare degree. I am satisfied that unusual success awaits it."
WILLIAM PEPPER, M. D.
Provost and Professor of Theory and Practice of Medicine and Clinical Medicine in the University of Pennsylvania.

"A clearly written, comprehensive manual. . . . One which we can commend to students as a reliable text-book, written with an evident knowledge of the wants of those entering upon the study of this special branch of medical science."—*British Medical Journal.*

"It is hardly too much to say that for the student and practitioner beginning the study of Ophthalmology, it is the best single volume at present published."—*Medical News.*

"It is a very useful, satisfactory, and safe guide for the student and the practitioner, and one of the best works of this scope in the English language."—*Annals of Ophthalmology.*

DISEASES OF WOMEN. By J. BLAND SUTTON, F. R. C. S., Assistant Surgeon to Middlesex Hospital, and Surgeon to Chelsea Hospital, London; and ARTHUR E. GILES, M. D., B. Sc., Lond., F. R. C. S., Edin., Assistant Surgeon to Chelsea Hospital, London. 436 pages, handsomely illustrated. Cloth, $2.50 net.

The authors have placed in the hands of the physician and student a concise yet comprehensive guide to the study of gynecology in its most modern development. It has been their aim to relate facts and describe methods belonging to the science and art of gynecology in a way that will prove useful to students for examination purposes, and which will also enable the general physician to practice this important department of surgery with advantage to his patients and with satisfaction to himself.

"The book is very well prepared, and is certain to be well received by the medical public."
—*British Medical Journal.*

"The text has been carefully prepared. Nothing essential has been omitted, and its teachings are those recommended by the leading authorities of the day."—*Journal of the American Medical Association.*

TEXT-BOOK UPON THE PATHOGENIC BACTERIA. Specially written for Students of Medicine. By JOSEPH MCFARLAND, M. D., Professor of Pathology and Bacteriology in the Medico-Chirurgical College of Philadelphia, etc. 497 pages, finely illustrated. Price, Cloth, $2.50 net.

SECOND EDITION, REVISED AND GREATLY ENLARGED.

The work is intended to be a text-book for the medical student and for the practitioner who has had no recent laboratory training in this department of medical science. The instructions given as to needed apparatus, cultures, stainings, microscopic examinations, etc. are ample for the student's needs, and will afford to the physician much information that will interest and profit him relative to a subject which modern science shows to go far in explaining the etiology of many diseased conditions.

In this second edition the work has been brought up to date in all departments of the subject, and numerous additions have been made to the technique in the endeavor to make the book fulfil the double purpose of a systematic work upon bacteria and a laboratory guide.

"It is excellently adapted for the medical students and practitioners for whom it is avowedly written. . . . The descriptions given are accurate and readable, and the book should prove useful to those for whom it is written.—*London Lancet*, Aug. 29, 1896.

"The author has succeeded admirably in presenting the essential details of bacteriological technics, together with a judiciously chosen summary of our present knowledge of pathogenic bacteria. . . . The work, we think, should have a wide circulation among English-speaking students of medicine."—*N. Y. Medical Journal*, April 4, 1896.

"The book will be found of considerable use by medical men who have not had a special bacteriological training, and who desire to understand this important branch of medical science."—*Edinburgh Medical Journal*, July, 1896.

LABORATORY GUIDE FOR THE BACTERIOLOGIST. By LANGDON FROTHINGHAM, M. D. V., Assistant in Bacteriology and Veterinary Science, Sheffield Scientific School, Yale University. Illustrated. Price, Cloth, 75 cents.

The technical methods involved in bacteria-culture, methods of staining, and microscopical study are fully described and arranged as simply and concisely as possible. The book is especially intended for use in laboratory work.

"It is a convenient and useful little work, and will more than repay the outlay necessary for its purchase in the saving of time which would otherwise be consumed in looking up the various points of technique so clearly and concisely laid down in its pages."—*American Med.-Surg. Bulletin.*

FEEDING IN EARLY INFANCY. By ARTHUR V. MEIGS, M. D. Bound in limp cloth, flush edges. Price, 25 cents net.

SYNOPSIS: Analyses of Milk—Importance of the Subject of Feeding in Early Infancy—Proportion of Casein and Sugar in Human Milk—Time to Begin Artificial Feeding of Infants—Amount of Food to be Administered at Each Feeding—Intervals between Feedings—Increase in Amount of Food at Different Periods of Infant Development—Unsuitableness of Condensed Milk as a Substitute for Mother's Milk—Objections to Sterilization or "Pasteurization" of Milk—Advances made in the Method of Artificial Feeding of Infants.

MATERIA MEDICA FOR NURSES. By EMILY A. M. STONEY, Graduate of the Training-school for Nurses, Lawrence, Mass.; late Superintendent of the Training-school for Nurses, Carney Hospital, South Boston, Mass. Handsome octavo, 300 pages. Cloth, $1.50 net.

The present book differs from other similar works in several features, all of which are introduced to render it more practical and generally useful. The general plan of contents follows the lines laid down in training-schools for nurses, but the book contains much useful matter not usually included in works of this character, such as Poison-emergencies, Ready Dose-list, Weights and Measures, etc., as well as a Glossary, defining all the terms in Materia Medica, and describing all the latest drugs and remedies, which have been generally neglected by other books of the kind.

ESSENTIALS OF ANATOMY AND MANUAL OF PRACTICAL DISSECTION, containing "Hints on Dissection." By CHARLES B. NANCREDE, M. D., Professor of Surgery and Clinical Surgery in the University of Michigan, Ann Arbor; Corresponding Member of the Royal Academy of Medicine, Rome, Italy; late Surgeon Jefferson Medical College, etc. Fourth and revised edition. Post 8vo, over 500 pages, with handsome full-page lithographic plates in colors, and over 200 illustrations. Price: Extra Cloth or Oilcloth for the dissection-room, $2.00 net.

Neither pains nor expense has been spared to make this work the most exhaustive yet concise Student's Manual of Anatomy and Dissection ever published, either in America or in Europe.

The colored plates are designed to aid the student in dissecting the muscles arteries, veins, and nerves. The wood-cuts have all been specially drawn and engraved, and an Appendix added containing 60 illustrations representing the structure of the entire human skeleton, the whole being based on the eleventh edition of Gray's *Anatomy*.

A MANUAL OF PRACTICE OF MEDICINE. By A. A. STEVENS, A. M., M. D., Instructor in Physical Diagnosis in the University of Pennsylvania, and Professor of Pathology in the Woman's Medical College of Pennsylvania. Specially intended for students preparing for graduation and hospital examinations. Post 8vo, 519 pages. Numerous illustrations and selected formulæ. Price, bound in flexible leather, $2.00 net.

FIFTH EDITION, REVISED AND ENLARGED.

Contributions to the science of medicine have poured in so rapidly during the last quarter of a century that it is well-nigh impossible for the student, with the limited time at his disposal, to master elaborate treatises or to cull from them that knowledge which is absolutely essential. From an extended experience in teaching, the author has been enabled, by classification, to group allied symptoms, and by the judicious elimination of theories and redundant explanations to bring within a comparatively small compass a complete outline of the practice of medicine.

MANUAL OF MATERIA MEDICA AND THERAPEUTICS.

By A. A. STEVENS, A. M., M. D., Instructor of Physical Diagnosis in the University of Pennsylvania, and Professor of Pathology in the Woman's Medical College of Pennsylvania. 445 pages. Price, bound in flexible leather, $2.25.

SECOND EDITION, REVISED.

This wholly new volume, which is based on the last edition of the *Pharmacopœia*, comprehends the following sections: Physiological Action of Drugs; Drugs; Remedial Measures other than Drugs; Applied Therapeutics; Incompatibility in Prescriptions; Table of Doses; Index of Drugs; and Index of Diseases; the treatment being elucidated by more than two hundred formulæ.

"The author is to be congratulated upon having presented the medical student with as accurate a manual of therapeutics as it is possible to prepare."—*Therapeutic Gazette.*

"Far superior to most of its class; in fact, it is very good. Moreover, the book is reliable and accurate."—*New York Medical Journal.*

"The author has faithfully presented modern therapeutics in a comprehensive work, . . . and it will be found a reliable guide."—*University Medical Magazine.*

NOTES ON THE NEWER REMEDIES: their Therapeutic Applications and Modes of Administration.

By DAVID CERNA, M. D., PH. D., Demonstrator of and Lecturer on Experimental Therapeutics in the University of Pennsylvania. Post-octavo, 253 pages. Price, $1.25.

SECOND EDITION, RE-WRITTEN AND GREATLY ENLARGED.

The work takes up in alphabetical order all the newer remedies, giving their physical properties, solubility, therapeutic applications, administration, and chemical formula.

It thus forms a very valuable addition to the various works on therapeutics now in existence.

Chemists are so multiplying compounds, that, if each compound is to be thoroughly studied, investigations must be carried far enough to determine the practical importance of the new agents.

"Especially valuable because of its completeness, its accuracy, its systematic consideration of the properties and therapy of many remedies of which doctors generally know but little, expressed in a brief yet terse manner."—*Chicago Clinical Review.*

TEMPERATURE CHART.

Prepared by D. T. LAINÉ, M. D. Size 8 x 13½ inches. Price, per pad of 25 charts, 50 cents.

A conveniently arranged chart for recording Temperature, with columns for daily amounts of Urinary and Fecal Excretions, Food, Remarks, etc. On the back of each chart is given in full the method of Brand in the treatment of Typhoid Fever.

A TEXT-BOOK OF HISTOLOGY, DESCRIPTIVE AND PRACTICAL. For the Use of Students. By ARTHUR CLARKSON, M. B., C. M., Edin., formerly Demonstrator of Physiology in the Owen's College, Manchester; late Demonstrator of Physiology in the Yorkshire College, Leeds. Large 8vo, 554 pages, with 22 engravings in the text, and 174 beautifully colored original illustrations. Price, strongly bound in Cloth, $4.00 net.

The purpose of the writer in this work has been to furnish the student of Histology, in one volume, with both the descriptive and the practical part of the science. The first two chapters are devoted to the consideration of the general methods of Histology; subsequently, in each chapter, the structure of the tissue or organ is first systematically described, the student is then taken tutorially over the specimens illustrating it, and, finally, an appendix affords a short note of the methods of preparation.

"The work must be considered a valuable addition to the list of available text-books, and is to be highly recommended."—*New York Medical Journal.*

"One of the best works for students we have ever noticed. We predict that the book will attain a well-deserved popularity among our students."—*Chicago Medical Recorder.*

THE PATHOLOGY AND TREATMENT OF SEXUAL IMPOTENCE. By VICTOR G. VECKI, M. D. From the second German edition, revised and rewritten. Demi-octavo, about 300 pages. Cloth, $2.00 net.

The subject of impotence has but seldom been treated in this country in the truly scientific spirit that it deserves, and this volume will come to many as a revelation of the possibilities of therapeusis in this important field. Dr. Vecki's work has long been favorably known, and the German book has received the highest consideration. This edition is more than a mere translation, for, although based on the German edition, it has been entirely rewritten by the author in English.

"The work can be recommended as a scholarly treatise on its subject, and it can be read with advantage by many practitioners."—*Journal of the American Medical Association.*

THE TREATMENT OF PELVIC INFLAMMATIONS THROUGH THE VAGINA. By W. R. PRYOR, M. D., Professor of Gynecology in the New York Polyclinic. 12mo, 248 pages, handsomely illustrated. Cloth, $2.00 net.

In this book the author directs the attention of the general practitioner to a surgical treatment of the pelvic diseases of women. There exists the utmost confusion in the profession regarding the most successful methods of treating pelvic inflammations; and inasmuch as inflammatory lesions constitute the majority of all pelvic diseases, the subject is an important one. It has been the endeavor of the author to put down every little detail, no matter how insignificant, which might be of service.

DISEASES OF WOMEN. By HENRY J. GARRIGUES, A.M., M.D., Professor of Gynecology in the New York School of Clinical Medicine; Gynecologist to St. Mark's Hospital and to the German Dispensary, New York City. In one handsome octavo volume of 728 pages, illustrated by 335 engravings and colored plates. Prices: Cloth, $4.00 net; Sheep or Half-Morocco, $5.00 net.

A PRACTICAL work on gynecology for the use of students and practitioners, written in a terse and concise manner. The importance of a thorough knowledge of the anatomy of the female pelvic organs has been fully recognized by the author, and considerable space has been devoted to the subject. The chapters on Operations and on Treatment are thoroughly modern, and are based upon the large hospital and private practice of the author. The text is elucidated by a large number of illustrations and colored plates, many of them being original, and forming a complete atlas for studying *embryology* and the *anatomy* of the *female genitalia*, besides exemplifying, whenever needed, morbid conditions, instruments, apparatus, and operations.

Second Edition, Thoroughly Revised.

The first edition of this work met with a most appreciative reception by the medical press and profession both in this country and abroad, and was adopted as a text-book or recommended as a book of reference by nearly *one hundred* colleges in the United States and Canada. The author has availed himself of the opportunity afforded by this revision to embody the latest approved advances in the treatment employed in this important branch of Medicine. He has also more extensively expressed his own opinion on the comparative value of the different methods of treatment employed.

"One of the best text-books for students and practitioners which has been published in the English language; it is condensed, clear, and comprehensive. The profound learning and great clinical experience of the distinguished author find expression in this book in a most attractive and instructive form. Young practitioners, to whom experienced consultants may not be available, will find in this book invaluable counsel and help."

THAD. A. REAMY, M.D., LL.D.,
Professor of Clinical Gynecology, Medical College of Ohio; Gynecologist to the Good Samaritan and Cincinnati Hospitals.

A SYLLABUS OF GYNECOLOGY, arranged in conformity with "An American Text-Book of Gynecology." By J. W. LONG, M.D., Professor of Diseases of Women and Children, Medical College of Virginia, etc. Price, Cloth (interleaved), $1.00 net.

Based upon the teaching and methods laid down in the larger work, this will not only be useful as a supplementary volume, but to those who do not already possess the text-book it will also have an independent value as an aid to the practitioner in gynecological work, and to the student as a guide in the lecture-room, as the subject is presented in a manner at once systematic, clear, succinct, and practical.

THE AMERICAN POCKET MEDICAL DICTIONARY. Edited by W. A. NEWMAN DORLAND, M. D., Assistant Obstetrician to the Hospital of the University of Pennsylvania; Fellow of the American Academy of Medicine. Containing the pronunciation and definition of all the principal words used in medicine and the kindred sciences, with 64 extensive tables. Handsomely bound in flexible leather, limp, with gold edges and patent thumb index. Price, $1.00 net; with thumb index, $1.25 net.

SECOND EDITION, REVISED.

This is the ideal pocket lexicon. It is an absolutely new book, and not a revision of any old work. It is complete, defining all the terms of modern medicine and forming an unusually complete vocabulary. It gives the pronunciation of all the terms. It makes a special feature of the newer words neglected by other dictionaries. It contains a wealth of anatomical tables of special value to students. It forms a handy volume, indispensable to every medical man.

SAUNDERS' POCKET MEDICAL FORMULARY. By WILLIAM M. POWELL, M. D., Attending Physician to the Mercer House for Invalid Women at Atlantic City. Containing 1800 Formulæ, selected from several hundred of the best-known authorities. Forming a handsome and convenient pocket companion of nearly 300 printed pages, with blank leaves for Additions; with an Appendix containing Posological Table, Formulæ and Doses for Hypodermatic Medication, Poisons and their Antidotes, Diameters of the Female Pelvis and Fœtal Head, Obstetrical Table, Diet List for Various Diseases, Materials and Drugs used in Antiseptic Surgery, Treatment of Asphyxia from Drowning, Surgical Remembrancer, Tables of Incompatibles, Eruptive Fevers, Weights and Measures, etc. Handsomely bound in morocco, with side index, wallet, and flap. Price, $1.75 net.

FIFTH EDITION, THOROUGHLY REVISED.

"This little book, that can be conveniently carried in the pocket, contains an immense amount of material. It is very useful, and as the name of the author of each prescription is given, is unusually reliable."—*New York Medical Record.*

A COMPENDIUM OF INSANITY. By JOHN B. CHAPIN, M.D., LL.D., Physician-in-Chief, Pennsylvania Hospital for the Insane; late Physician-Superintendent of the Willard State Hospital, New York; Honorary Member of the Medico-Psychological Society of Great Britain, of the Society of Mental Medicine of Belgium. 12mo, 234 pages, illust. Cloth, $1.25 net.

The author has given, in a condensed and concise form, a compendium of Diseases of the Mind, for the convenient use and aid of physicians and students. It contains a clear, concise statement of the clinical aspects of the various abnormal mental conditions, with directions as to the most approved methods of managing and treating the insane.

"The practical parts of Dr. Chapin's book are what constitute its distinctive merit. We desire especially, however, to call attention to the fact that in the subject of the therapeutics of insanity the work is exceedingly valuable. The author has made a distinct addition to the literature of his specialty."—*Philadelphia Medical Journal.*

AN OPERATION BLANK, with Lists of Instruments, etc. required in Various Operations. Prepared by W. W. KEEN, M. D., LL.D., Professor of Principles of Surgery in the Jefferson Medical College, Philadelphia. Price per Pad, containing Blanks for fifty operations, 50 cents net.

SECOND EDITION, REVISED FORM.

A convenient blank, suitable for all operations, giving complete instructions regarding necessary preparation of patient, etc., with a full list of dressings and medicines to be employed.

On the back of each blank is a list of instruments used—viz. general instruments, etc., required for all operations; and special instruments for surgery of the brain and spine, mouth and throat, abdomen, rectum, male and female genito-urinary organs, the bones, etc.

The whole forming a neat pad, arranged for hanging on the wall of a surgeon's office or in the hospital operating-room.

"Will serve a useful purpose for the surgeon in reminding him of the details of preparation for the patient and the room as well as for the instruments, dressings, and antiseptics needed "—*New York Medical Record*

"Covers about all that can be needed in any operation."—*American Lancet.*

"The plan is a capital one."—*Boston Medical and Surgical Journal.*

LABORATORY EXERCISES IN BOTANY. By EDSON S. BASTIN, M. A., Professor of Materia Medica and Botany in the Philadelphia College of Pharmacy. Octavo volume of 536 pages, 87 full-page plates. Price, Cloth, $2.50.

This work is intended for the beginner and the advanced student, and it fully covers the structure of flowering plants, roots, ordinary stems, rhizomes, tubers, bulbs, leaves, flowers, fruits, and seeds. Particular attention is given to the gross and microscopical structure of plants, and to those used in medicine. Illustrations have freely been used to elucidate the text, and a complete index to facilitate reference has been added.

"There is no work like it in the pharmaceutical or botanical literature of this country, and we predict for it a wide circulation."—*American Journal of Pharmacy.*

DIET IN SICKNESS AND IN HEALTH. By MRS. ERNEST HART, formerly Student of the Faculty of Medicine of Paris and of the London School of Medicine for Women; with an INTRODUCTION by Sir Henry Thompson, F. R. C. S., M. D., London. 220 pages; illustrated. Price, Cloth, $1.50.

Useful to those who have to nurse, feed, and prescribe for the sick. In each case the accepted causation of the disease and the reasons for the special diet prescribed are briefly described. Medical men will find the dietaries and recipes practically useful, and likely to save them trouble in directing the dietetic treatment of patients.

A MANUAL OF PHYSIOLOGY, with Practical Exercises. For Students and Practitioners. By G. N. STEWART, M. A., M. D., D. Sc., lately Examiner in Physiology, University of Aberdeen, and of the New Museums, Cambridge University; Professor of Physiology in the Western Reserve University, Cleveland, Ohio. Handsome octavo volume of 848 pages, with 300 illustrations in the text, and 5 colored plates. Price, Cloth, $3.75 net.

THIRD EDITION, REVISED.

"It will make its way by sheer force of merit, and *amply deserves to do so. It is one of the very best English text-books on the subject.*"—*London Lancet.*

"Of the many text-books of physiology published, we do not know of one that so nearly comes up to the ideal as does Professor Stewart's volume."—*British Medical Journal.*

ESSENTIALS OF PHYSICAL DIAGNOSIS OF THE THORAX. By ARTHUR M. CORWIN, A. M., M. D., Demonstrator of Physical Diagnosis in the Rush Medical College, Chicago; Attending Physician to the Central Free Dispensary, Department of Rhinology, Laryngology, and Diseases of the Chest. 219 pages. Illustrated. Cloth, flexible covers. Price, $1.25 net.

THIRD EDITION, THOROUGHLY REVISED AND ENLARGED.

SYLLABUS OF OBSTETRICAL LECTURES in the Medical Department, University of Pennsylvania. By RICHARD C. NORRIS, A. M., M. D., Lecturer on Clinical and Operative Obstetrics, University of Pennsylvania. Third edition, thoroughly revised and enlarged. Crown 8vo. Price, Cloth, interleaved for notes, $2.00 net.

"This work is so far superior to others on the same subject that we take pleasure in calling attention briefly to its excellent features. It covers the subject thoroughly, and will prove invaluable both to the student and the practitioner. The author has introduced a number of valuable hints which would only occur to one who was himself an experienced teacher of obstetrics. The subject-matter is clear, forcible, and modern. We are especially pleased with the portion devoted to the practical duties of the accoucheur, care of the child, etc. The paragraphs on antiseptics are admirable; there is no doubtful tone in the directions given. No details are regarded as unimportant; no minor matters omitted. We venture to say that even the old practitioner will find useful hints in this direction which he cannot afford to despise."—*New York Medical Record.*

A SYLLABUS OF LECTURES ON THE PRACTICE OF SURGERY, arranged in conformity with "An American Text-Book of Surgery." By N. SENN, M. D., PH. D., Professor of Surgery in Rush Medical College, Chicago, and in the Chicago Polyclinic. Price, $2.00.

This work by so eminent an author, himself one of the contributors to "An American Text-Book of Surgery," will prove of exceptional value to the advanced student who has adopted that work as his text-book. It is not only the syllabus of an unrivalled course of surgical practice, but it is also an epitome of or supplement to the larger work.

"The author has evidently spared no pains in making his Syllabus thoroughly comprehensive, and has added new matter and alluded to the most recent authors and operations. Full references are also given to all requisite details of surgical anatomy and pathology."—*British Medical Journal,* London.

THE CARE OF THE BABY. By J. P. CROZER GRIFFITH, M. D., Clinical Professor of Diseases of Children, University of Pennsylvania; Physician to the Children's Hospital, Philadelphia, etc. 404 pages, with 67 illustrations in the text, and 5 plates. 12mo. Price, $1.50.

SECOND EDITION, REVISED.

A reliable guide not only for mothers, but also for medical students and practitioners whose opportunities for observing children have been limited.

"The whole book is characterized by rare good sense, and is evidently written by a master hand. It can be read with benefit not only by mothers, but by medical students and by any practitioners who have not had large opportunities for observing children."—*American Journal of Obstetrics.*

THE NURSE'S DICTIONARY of Medical Terms and Nursing Treatment, containing Definitions of the Principal Medical and Nursing Terms, Abbreviations, and Physiological Names, and Descriptions of the Instruments, Drugs, Diseases, Accidents, Treatments, Operations, Foods, Appliances, etc. encountered in the ward or the sick-room. By HONNOR MORTEN, author of "How to Become a Nurse," "Sketches of Hospital Life," etc. 16mo, 140 pages. Price, Cloth, $1.00.

This little volume is intended for use merely as a small reference-book which can be consulted at the bedside or in the ward. It gives sufficient explanation to the nurse to enable her to comprehend a case until she has leisure to look up larger and fuller works on the subject.

DIET LISTS AND SICK-ROOM DIETARY. By JEROME B. THOMAS, M. D., Visiting Physician to the Home for Friendless Women and Children and to the Newsboys' Home; Assistant Visiting Physician to the Kings County Hospital; Assistant Bacteriologist, Brooklyn Health Department. Price, Cloth, $1.50 (Send for specimen List.)

One hundred and sixty detachable (perforated) diet lists for Albuminuria, Anæmia and Debility, Constipation, Diabetes, Diarrhœa, Dyspepsia, Fevers, Gout or Uric-Acid Diathesis, Obesity, and Tuberculosis. Also forty detachable sheets of Sick-Room Dietary, containing full instructions for preparation of easily-digested foods necessary for invalids. Each list is *numbered only*, the disease for which it is to be used in no case being mentioned, an index key being reserved for the physician's private use.

DIETS FOR INFANTS AND CHILDREN IN HEALTH AND IN DISEASE. By LOUIS STARR, M. D., Editor of "An American Text-Book of the Diseases of Children." 230 blanks (pocket-book size), perforated and neatly bound in flexible morocco. Price, $1.25 net.

The first series of blanks are prepared for the first seven months of infant life; each blank indicates the ingredients, but not the *quantities*, of the food, the latter directions being left for the physician. After the seventh month, modifications being less necessary, the diet lists are printed in full. *Formula* for the preparation of diluents and foods are appended.

CATALOGUE OF MEDICAL WORKS. 39

HOW TO EXAMINE FOR LIFE INSURANCE. By JOHN M. KEATING, M. D., Fellow of the College of Physicians and Surgeons of Philadelphia; Vice-President of the American Pædiatric Society; Ex-President of the Association of Life Insurance Medical Directors. Royal 8vo, 211 pages, with two large half-tone illustrations, and a plate prepared by Dr. McClellan from special dissections; also, numerous cuts to elucidate the text. Third edition. Price, Cloth, $2.00 net.

"This is by far the most useful book which has yet appeared on insurance examination, a subject of growing interest and importance. Not the least valuable portion of the volume is Part II., which consists of instructions issued to their examining physicians by twenty-four representative companies of this country. As the proofs of these instructions were corrected by the directors of the companies, they form the latest instructions obtainable. If for these alone, the book should be at the right hand of every physician interested in this special branch of medical science."—*The Medical News*, Philadelphia.

NURSING: ITS PRINCIPLES AND PRACTICE. By ISABEL ADAMS HAMPTON, Graduate of the New York Training School for Nurses attached to Bellevue Hospital; Superintendent of Nurses and Principal of the Training School for Nurses, Johns Hopkins Hospital, Baltimore, Md.; late Superintendent of Nurses, Illinois Training School for Nurses, Chicago, Ill. In one very handsome 12mo volume of 512 pages, illustrated. Price, Cloth, $2.00 net.

SECOND EDITION, REVISED AND ENLARGED.

This original work on the important subject of nursing is at once comprehensive and systematic. It is written in a clear, accurate, and readable style, suitable alike to the student and the lay reader. Such a work has long been a desideratum with those entrusted with the management of hospitals and the instruction of nurses in training-schools. It is also of especial value to the graduated nurse who desires to acquire a practical working knowledge of the care of the sick and the hygiene of the sick-room.

OBSTETRIC ACCIDENTS, EMERGENCIES, AND OPERATIONS. By L. CH. BOISLINIERE, M. D., late Emeritus Professor of Obstetrics in the St. Louis Medical College. 381 pages, handsomely illustrated. Price, $2.00 net.

"For the use of the practitioner who, when away from home, has not the opportunity of consulting a library or of calling a friend in consultation. He then, being thrown upon his own resources, will find this book of benefit in guiding and assisting him in emergencies."

INFANT'S WEIGHT CHART. Designed by J. P. CROZER GRIFFITH, M. D., Clinical Professor of Diseases of Children in the University of Pennsylvania. 25 charts in each pad. Price per pad, 50 cents net.

A convenient blank for keeping a record of the child's weight during the first two years of life. Printed on each chart is a curve representing the average weight of a healthy infant, so that any deviation from the normal can readily be detected.

SAUNDERS' NEW SERIES OF MANUALS

for Students and Practitioners.

THAT there exists a need for thoroughly reliable hand-books on the leading branches of Medicine and Surgery is a fact amply demonstrated by the favor with which the SAUNDERS NEW SERIES OF MANUALS have been received by medical students and practitioners and by the Medical Press. These manuals are not merely condensations from present literature, but are ably written by well-known authors and practitioners, most of them being teachers in representative American colleges. Each volume is concisely and authoritatively written and exhaustive in detail, without being encumbered with the introduction of "cases," which so largely expand the ordinary text-book. These manuals will therefore form an admirable collection of advanced lectures, useful alike to the medical student and the practitioner: to the latter, too busy to search through page after page of elaborate treatises for what he wants to know, they will prove of inestimable value; to the former they will afford safe guides to the essential points of study.

The SAUNDERS NEW SERIES OF MANUALS are conceded to be superior to any similar books now on the market. No other manuals afford so much information in such a concise and available form. A liberal expenditure has enabled the publisher to render the mechanical portion of the work worthy of the high literary standard attained by these books.

Any of these Manuals will be mailed on receipt of price (see next page for List).

SAUNDERS' NEW SERIES OF MANUALS.

VOLUMES PUBLISHED.

PHYSIOLOGY. By JOSEPH HOWARD RAYMOND, A. M., M. D., Professor of Physiology and Hygiene and Lecturer on Gynecology in the Long Island College Hospital, etc. Price, $1.25 net.

SURGERY, General and Operative. By JOHN CHALMERS DACOSTA, M. D., Professor of Practice of Surgery and Clinical Surgery, Jefferson Medical College, Philadelphia. Second edition, revised and greatly enlarged. Octavo, 911 pages, 386 illustrations. Cloth, $4.00 net; Half-Morocco, $5.00 net.

DOSE-BOOK AND MANUAL OF PRESCRIPTION-WRITING. By E. Q. THORNTON, M. D., Demonstrator of Therapeutics, Jefferson Medical College, Philadelphia. Price, $1.25 net.

MEDICAL JURISPRUDENCE. By HENRY C. CHAPMAN, M. D., Professor of Institutes of Medicine and Medical Jurisprudence in the Jefferson Medical College of Philadelphia, etc. Price, $1.50 net.

SURGICAL ASEPSIS. By CARL BECK, M. D., Surgeon to St. Mark's Hospital and to the German Poliklinik; Instructor in Surgery, New York Post-Graduate Medical School, etc. Price, $1.25 net.

MANUAL OF ANATOMY. By IRVING S. HAYNES, M. D., Adjunct Professor of Anatomy and Demonstrator of Anatomy, Medical Department of the New York University, etc. Price, $2.50 net.

SYPHILIS AND THE VENEREAL DISEASES. By JAMES NEVINS HYDE, M. D., Professor of Skin and Venereal Diseases, and FRANK H. MONTGOMERY, M. D., Lecturer on Dermatology and Genito-urinary Diseases in Rush Medical College, Chicago. Price, $2.50 net.

PRACTICE OF MEDICINE. By GEORGE ROE LOCKWOOD, M. D., Professor of Practice in the Woman's Medical College of the New York Infirmary, etc. Price, $2.50 net.

OBSTETRICS. By W. A. NEWMAN DORLAND, M. D., Assistant Demonstrator of Obstetrics, University of Pennsylvania; Chief of Gynecological Dispensary, Pennsylvania Hospital. Price, $2.50 net.

DISEASES OF WOMEN. By J. BLAND SUTTON, F. R. C. S., Assistant Surgeon to the Middlesex Hospital, and Surgeon to the Chelsea Hospital for Women, London; and ARTHUR E. GILES, M. D., B. Sc. Lond., F. R. C. S. Edin., Assistant Surgeon to the Chelsea Hospital for Women, London. 436 pages, handsomely illustrated. Price, $2.50 net.

IN PREPARATION.

NERVOUS DISEASES. By CHARLES W. BURR, M. D., Clinical Professor of Nervous Diseases, Medico-Chirurgical College, Philadelphia, etc.

*** There will be published in the same series, at short intervals, carefully prepared works on various subjects, by prominent specialists.

SAUNDERS' QUESTION COMPENDS.

Arranged in Question and Answer Form.

THE LATEST, MOST COMPLETE, and BEST ILLUSTRATED SERIES OF COMPENDS EVER ISSUED.

Now the Standard Authorities in Medical Literature

WITH

Students and Practitioners in every City of the United States and Canada.

THE REASON WHY.

They are the advance guard of "Student's Helps"—that DO HELP; they are the leaders in their special line, *well and authoritatively written by able men, who, as teachers in the large colleges, know exactly what is wanted by a student preparing for his examinations.* The judgment exercised in the selection of authors is fully demonstrated by their professional elevation. Chosen from the ranks of Demonstrators, Quiz-masters, and Assistants, most of them have become Professors and Lecturers in their respective colleges.

Each book is of convenient size (5 × 7 inches), containing on an average 250 pages, profusely illustrated, and elegantly printed in clear, readable type, on fine paper.

The entire series, numbering twenty-four subjects, has been kept thoroughly revised and enlarged when necessary, many of them being in their fourth and fifth editions.

TO SUM UP.

Although there are numerous other Quizzes, Manuals, Aids, etc. in the market, none of them approach the "Blue Series of Question Compends;" and the claim is made for the following points of excellence:

1. Professional distinction and reputation of authors.
2. Conciseness, clearness, and soundness of treatment.
3. Size of type and quality of paper and binding.

*** Any of these Compends will be mailed on receipt of price (see next page for List).

SAUNDERS' QUESTION-COMPEND SERIES.

Price, Cloth, $1.00 per copy, except when otherwise noted.

1. **ESSENTIALS OF PHYSIOLOGY.** 4th edition. Illustrated. Revised and enlarged. By H. A. Hare, M. D. (Price, $1.00 net.)
2. **ESSENTIALS OF SURGERY.** 7th edition, with a chapter on Appendicitis. 90 illustrations. By Edward Martin, M. D. (Price, $1.00 net.)
3. **ESSENTIALS OF ANATOMY.** 6th edition, thoroughly revised. 151 illustrations. By Charles B. Nancrede, M. D. (Price, $1.00 net.)
4. **ESSENTIALS OF MEDICAL CHEMISTRY, ORGANIC AND INORGANIC.** 5th edition, revised, with an Appendix. By Lawrence Wolff, M. D. ($1.00 net.)
5. **ESSENTIALS OF OBSTETRICS.** 4th edition, revised and enlarged. 75 illustrations. By W. Easterly Ashton, M. D.
6. **ESSENTIALS OF PATHOLOGY AND MORBID ANATOMY.** 7th thousand. 46 illustrations. By C. E. Armand Semple, M. D.
7. **ESSENTIALS OF MATERIA MEDICA, THERAPEUTICS, AND PRESCRIPTION-WRITING.** 5th edition. By Henry Morris, M. D.
8, 9. **ESSENTIALS OF PRACTICE OF MEDICINE.** By Henry Morris, M. D. An Appendix on Urine Examination. Illustrated. By Lawrence Wolff, M. D. 3d edition, enlarged by some 300 Essential Formulæ, selected from eminent authorities, by Wm. M. Powell, M. D. (Double number, price $2.00.)
10. **ESSENTIALS OF GYNÆCOLOGY.** 4th edition, revised. With 62 illustrations. By Edwin B. Cragin, M. D.
11. **ESSENTIALS OF DISEASES OF THE SKIN.** 4th edition, revised and enlarged. 71 letter-press cuts and 15 half-tone illustrations. By Henry W. Stelwagon, M.D. (Price, $1.00 net.)
12. **ESSENTIALS OF MINOR SURGERY, BANDAGING, AND VENEREAL DISEASES.** 2d edition, revised and enlarged. 78 illustrations. By Edward Martin, M. D.
13. **ESSENTIALS OF LEGAL MEDICINE, TOXICOLOGY, AND HYGIENE.** 130 illustrations. By C. E. Armand Semple, M. D.
14. **ESSENTIALS OF DISEASES OF THE EYE, NOSE, AND THROAT.** 124 illustrations. 2d edition, revised. By Edward Jackson, M. D., and E. Baldwin Gleason, M. D.
15. **ESSENTIALS OF DISEASES OF CHILDREN.** 2d edition. By William M. Powell, M. D.
16. **ESSENTIALS OF EXAMINATION OF URINE.** Colored "Vogel Scale," and numerous illustrations. By Lawrence Wolff, M. D. (Price, 75 cents.)
17. **ESSENTIALS OF DIAGNOSIS.** 2d edition, thoroughly revised. 60 illustrations. By S. Solis-Cohen, M. D., and A. A. Eshner, M. D. (Price, $1.00 net.)
18. **ESSENTIALS OF PRACTICE OF PHARMACY.** 2d edition, revised. By L. E. Sayre.
20. **ESSENTIALS OF BACTERIOLOGY.** 3d edition. 82 illustrations. By M. V. Ball, M. D.
21. **ESSENTIALS OF NERVOUS DISEASES AND INSANITY.** 48 illustrations. 3d edition, revised. By John C. Shaw, M. D.
22. **ESSENTIALS OF MEDICAL PHYSICS.** 155 illustrations. 2d edition, revised. By Fred J. Brockway, M. D. (Price, $1.00 net.)
23. **ESSENTIALS OF MEDICAL ELECTRICITY.** 65 illustrations. By David D. Stewart, M. D., and Edward S. Lawrance, M. D.
24. **ESSENTIALS OF DISEASES OF THE EAR.** 114 illustrations. 2d edition, revised and enlarged. By E. Baldwin Gleason, M. D.

Some of the Books in Preparation for Publication during 1900.

AMERICAN Text=Book of Pathology.
Edited by LUDVIG HEKTOEN, M.D., Professor of Pathology, Rush Medical College, Chicago; and DAVID RIESMAN, M.D., Demonstrator of Pathological Histology, University of Pennsylvania.

AMERICAN Text=Book of Legal Medicine and Toxicology.
Edited by FREDERICK PETERSON, M.D., Chief of Clinic, Nervous Department, College of Physicians and Surgeons, New York City; and WALTER S. HAINES, M.D., Professor of Chemistry, Pharmacy, and Toxicology, Rush Medical College, Chicago.

BECK—Fractures.
By CARL BECK, M.D., Professor of Surgery in the N. Y. School of Clinical Medicine.

BÖHM, DAVIDOFF, and HUBER—A Text=Book of Human Histology.
Including Microscopic Technic. By DR A A. BÖHM and DR. M. VON DAVIDOFF, of the Anatomical Institute of Munich, and G. C. HUBER, M.D., Junior Professor of Anatomy and Histology, University of Michigan, Ann Arbor.

EICHHORST—A Text=Book of the Practice of Medicine.
By DR. HERMAN EICHHORST, Professor of Special Pathology and Therapeutics and Director of the Medical Clinic, University of Zurich. Translated and edited by AUGUSTUS A. ESHNER, M.D, Professor of Clinical Medicine in the Philadelphia Polyclinic.

FRIEDRICH — Rhinology, Laryngology, and Otology in their Relations to General Medicine.
By DR. E. P. FRIEDRICH, of the University of Leipsig.

LEVY AND KLEMPERER — The Elements of Clinical Bacteriology.
By DR. ERNST LEVY, Professor in the University of Strassburg, and DR. FELIX KLEMPERER, Privat-Docent in the University of Strassburg. Translated and edited by AUGUSTUS A. ESHNER, M.D., Professor of Clinical Medicine in the Philadelphia Polyclinic. *Just Ready.* Cloth, $2.50 net.

McFARLAND—A Text=Book of Pathology.
By JOSEPH MCFARLAND, M.D., Professor of Pathology and Bacteriology, Medico-Chirurgical College, Philadelphia.

OGDEN — Clinical Examination of the Urine.
By J. BERGEN OGDEN, M.D., Assistant in Chemistry, Harvard Medical School.

PYLE—A Manual of Personal Hygiene.
Edited by WALTER L. PYLE, M.D., Assistant Surgeon to Wills' Eye Hospital, Philada.

SCUDDER—The Treatment of Fractures.
By CHARLES L. SCUDDER, M.D., Assistant in Clinical and Operative Surgery, Harvard University.

SENN—Practical Surgery.
By NICHOLAS SENN, M.D., PH.D., LL.D., Professor of the Practice of Surgery and of Clinical Surgery, Rush Medical College, Chicago. Octavo volume of about 800 pages, profusely illustrated.

The Pathology and Treatment of Tumors.
By NICHOLAS SENN, M.D., PH.D., LL.D., Professor of the Practice of Surgery and of Clinical Surgery, Rush Medical College, Chicago. A New and Thoroughly Revised Edition in preparation.

STENGEL AND WHITE — The Blood in its Clinical and Pathological Relations.
By ALFRED STENGEL, M.D., Professor of Clinical Medicine, University of Pennsylvania; and C. Y. WHITE, M.D., Instructor in Clinical Medicine, University of Pennsylvania.

STEVENS—The Physical Diagnosis of Diseases of the Chest.
By A. A. STEVENS, A.M., M.D., Lecturer on Terminology, and Instructor in Physical Diagnosis, University of Pennsylvania.

STONEY — Surgical Technique for Nurses.
By EMILY A. M. STONEY, late Superintendent of the Training Schools for Nurses, Carney Hospital, South Boston, Mass.

www.ingramcontent.com/pod-product-compliance
Lightning Source LLC
Chambersburg PA
CBHW032058220426
43664CB00008B/1047